Springer Series on Bio- and Neurosystems

Volume 17

The Springer Series on Bio- and Neurosystems publishes fundamental principles and state-of-the-art research at the intersection of biology, neuroscience, information processing and the engineering sciences. The series covers general informatics methods and techniques, together with their use to answer biological or medical questions. Of interest are both basics and new developments on traditional methods such as machine learning, artificial neural networks, statistical methods, nonlinear dynamics, information processing methods, and image and signal processing. New findings in biology and neuroscience obtained through informatics and engineering methods, topics in systems biology, medicine, neuroscience and ecology, as well as engineering applications such as robotic rehabilitation, health information technologies, and many more, are also examined. The main target group includes informaticians and engineers interested in biology, neuroscience and medicine, as well as biologists and neuroscientists using computational and engineering tools. Volumes published in the series include monographs, edited volumes, and selected conference proceedings. Books purposely devoted to supporting education at the graduate and post-graduate levels in bio- and neuroinformatics, computational biology and neuroscience, systems biology, systems neuroscience and other related areas are of particular interest.

All books published in the series are submitted for consideration in Web of Science.

Bernard Widrow · Edward P. Katz

Cognitive Memory

Human Memory | Machine Memory

Springer

Bernard Widrow
Stanford University
Stanford, CA, USA

Edward P. Katz
Stanford University
Stanford, CA, USA

ISSN 2520-8535 ISSN 2520-8543 (electronic)
Springer Series on Bio- and Neurosystems
ISBN 978-3-031-80938-5 ISBN 978-3-031-80939-2 (eBook)
https://doi.org/10.1007/978-3-031-80939-2

This Springer imprint is published by the registered company Springer Nature Switzerland AG
The registered company address is: Gewerbestrasse 11, 6330 Cham, Switzerland

To the memory of David Hubel
and to the memory of Béla Julesz

Foreword by Nikola K. Kasabov

Memory and learning are the very basic principles of human-, animal- and artificial intelligence. How memory is formed in the human brain and how we can model it in artificial systems? These are very old questions, and despite the fact that they have many answers so far, these questions are still challenging for both neuroscientists and computer scientists. And this book is a contribution to the continuous search for answers.

Memory in living organism is evolving, adapting through learning to make the organism adapt to the environment and to survive. The evolution/adaptation at an individual level and at a population/species level is still to be better understood as part of the progress in biology and neuroscience, but also for the sake of the creation of more intelligent, adaptive, life-long learning engineering systems.

And here comes this monograph book by the pioneer of artificial neural networks and engineering learning systems Bernie Widrow, co-authored with his colleague Edward Katz to cover in a systematic way various aspects of memory and its adaptivity in both living and engineering systems.

Widrow is Originator (with Hoff) of the Least Mean Square (LMS) learning algorithm for artificial neural networks (1960) implemented in electronic machines called ADALINE and MADALINE. And that is where the book starts. Widrow introduced the Hebbian-LMS (with Kim and Park, 2015) as a continuation of the Hebbian learning rule introduced by Donald Hebb in 1949 as a synaptic neuronal learning rule. The Hebbian-LMS algorithm can be used to model learning and adaptability at different levels.

The book covers a wide spectrum of material, from biological neurons, synapses and neuronal learning algorithms, DNA, to applications across neuroscience, psychology (cognition), artificial neural networks and their applications. As it is presented, memory is unsupervised and its acquisition is always "*gated*" by interest. And memory is always *associative*, so that retrieval can be achieved through presenting only partial input patterns.

The book explains many principles of human cognition and computational intelligence, such as learning, thinking, short-term memory, long-term memory, and decision-making with simple examples, e.g. explaining reinforcement learning with weeds taking over grass (Chap. 14). This is easy to explain even to children. The style of explaining complex concepts with simple phenomena in nature is prevailing in the book and that is what makes this book unique. And indeed, all we do is nature inspired, in one way or another, and it takes *wisdom* to explain it in this way.

A large number of applications of cognitive memory, both biological and artificial, are presented, including, but not limited to: image recognition; sound recognition; motor control; Alzheimer disease; post-traumatic stress disorder; dementia; and many others.

The book also presents some challenges to neuroscientists with the claim that neural networks are involved only in learning and retrieving the memory, but the memory resides in the DNA of the nucleus of cells and one of the arguments is that living neural networks change all the time or die (Chap. 16). This is in contrast to the common understanding that memory is contained in the synaptic connections of the biological and also the artificial neural networks, and of course, genetic memory is contained in the DNA.

Some philosophical aspects presented in the book are also challenging scientists, at least partially, e.g. "the meaning of things is what is remembered". All these challenges make the book very interesting to read and for readers to find some proofs of the claims in the book.

The book is a wonderful historic review of the evolution of concepts and human cognition and in intelligent system, rather than a presentation of the current state of AI. It is presented in a very readable form, with lots and lots of personal memories and stories of Widrow, such as photographs, interviews, personal contacts with prominent scientists, and personal experiences. And this is another peculiarity of the book, that makes it very special and suitable for a pioneering scientist such as Bernie Widrow.

Auckland, New Zealand Prof. Nikola K. Kasabov

Prof. Nikola K. Kasabov is Professor Emeritus of knowledge engineering and Founding Director of the KEDRI Institute at Auckland University of Technology (AUT), New Zealand. He is also Founding Director of Knowledgeengineering.ai. Kasabov holds a Master's degree in Computer Engineering and Ph.D. in Mathematics from TU Sofia, Bulgaria, and Doctor Honoris Causa from Obuda University, Hungary. He is Life Fellow of IEEE, Fellow of the Royal Society (Academy) of New Zealand, Fellow of the INNS College of Fellows. Kasabov is Visiting Professor at the IICT of the Bulgarian Academy of Sciences and Dalian University, China. He has published more than 700 works in the areas of neural networks, computational intelligence, evolving connectionist systems, neuro-informatics, bioinformatics. He is Past President of the International Neural Network Society (INNS) and Asia-Pacific Neural Network Society (APNNS). He is Series Editor of the *Springer Series on Bio- and Neurosystems*.

Foreword by Henry Yin

What is most surprising about Bernard Widrow's new book (coauthored with Edward Katz) is not the age of the senior author (ninety-four), but that it sheds light on memory, one of nature's remaining mysteries. It is based on his knowledge of computer memory, his experience as a researcher on neural networks, and his subjective experience as a user of human memory. Although Widrow's work has led him to think extensively about memory, he is Outsider to mainstream memory research. Indeed, this book challenges some of the most popular ideas on memory.

Widrow's first unconventional idea is about memory storage. According to him, cognitive memory stores inputs from the sensory system automatically, but this process is normally gated by how "interesting" the content is. In most humans, this gating appears to be stringent, so that only a very small portion gets stored in long-term memory. When such gating is disrupted (which happens rarely), the consequence is photographic memory. Béla Julesz, to whom this book is dedicated, performed one of the most striking experiments on photographic memory. Taking advantage of stereoscopic vision, Julesz could project two different polarized random-dot images simultaneously to both eyes. Even though each eye alone only sees random dots, with inputs from both eyes the brain could create a perception of three-dimensional objects. In an elegant experiment, he used a variant of this procedure on a girl with photographic memory. He first showed her a random dot pattern to her left eye, and a week later, he showed another random pattern to her right eye. Remarkably, she could see the pattern that is only available in the binocular image by combining current visual input from the right eye and the remembered random dot image viewed with the left eye.

The accurate stored representation of the random dot image in Julesz's experiment cannot be explained by common theories of memory. According to an influential view, memory is stored in the weights of a neural network, in the connections between neurons. This view was first proposed in the 19th century by Sherrington and Cajal and later made popular by Hebb and students of synaptic plasticity. Today, it is commonly assumed that, after training, the weights of a neural network collectively represent the stored memory, while the training patterns are discarded. Although Widrow himself is a pioneer in neural network research, he does not believe that these weights can provide a mechanism for storing long-term memory. Instead, he argues that memory storage takes place in the cell nucleus, probably in the DNA.

Although the idea that memory storage occurs in the DNA is not new, Widrow's proposal is innovative as he finds a different role for neural networks in memory. This brings us to his second idea—that neural networks are used primarily as the means for memory retrieval, not storage. From the very beginning, neural networks have been used for pattern recognition (Rosenblatt's Perceptron or Widrow's Adaline). Widrow argues persuasively that pattern recognition and memory retrieval involve the same underlying mechanism, namely the autoassociative neural network. Much of perception (and pattern recognition) requires a learning process which results in the formation of long-term memory templates. During memory retrieval, one uses current perceptual inputs (prompts) to retrieve specific pieces of previously stored memory. Recognition occurs when the input pattern fits stored patterns. If the patterns are similar, the autoassociative neural network connects the prompt patterns to patterns recorded in memory, generating memory retrieval.

In addition to presenting a serious challenge to the current paradigm in memory research, this book is also surprisingly personal. It is interspersed with more lighthearted vignettes, ranging from Widrow's experience with music and sculpture to encounters with famous scientists. These vignettes hint at a life richly lived and vividly remembered and shed light on the mind and memory of an intellectual pioneer whose work has changed the world.

Henry Yin, Ph.D.
Professor of Psychology
and Neuroscience
Professor of Neurobiology
Duke University
Durham, USA

Foreword by Greg Kovacs

It is an honor and a pleasure to write a foreword for this book.

I have truly enjoyed knowing Prof. Widrow for almost forty years. He took me under wing when I was a wet-behind-the-ears medical student around 1985. He mentored, encouraged, supported, and befriended me. Little did I know then that I was getting all of this kindness from one of the true icons of Silicon Valley and beyond.

Over the years, our relationship evolved and grew. I joined Stanford's EE faculty in 1991, and we collaborated on many research projects over the years. I'm very proud to have worked with him, and our mutual graduate students still speak very fondly of him. Our friendship grew, and in the end, Bernie is part of our family, with his kindness and mentorship extending to our two sons, one of whom, Reid, is quite literally following in Bernie's footsteps as he works toward his EE Ph.D. at MIT.

Even today, I get mentorship, we collaborate, and we are great friends. I continue to be amazed that at the tender young age of 94, Bernie is still sharp, witty, never forgets his famous stories, and can write a fascinating book.

This book, his sixth if my memory serves me well, is in keeping with his latest, *Cybernetics 2.0*, in its warm style, filled with stories and real-world examples. Here,

Bernie explores consciousness, memory and the parallels between adaptive neural networks and the human mind. From the perspective of a brilliant engineer he brings the reader many interesting and relatable examples. For those schooled in classical neuroscience, some concepts may be controversial, but nonetheless, fascinating. All comers to this book will learn and find delightful nuggets of thought.

Having said that, I will close my foreword with heartfelt thanks to Bernie Widrow. Your friendship over these many years has been life-changing in a very positive way for myself, and the countless others you have influenced. You are a hero to us all.

Greg Kovacs, MD, Ph.D.
Fellow, IEEE, AIMBE, Emeritus
Professor of Electrical Engineering
Stanford University
Stanford, USA

Preface

A cognitive memory is defined here as a human memory. Cognitive memory can also refer to an artificial memory that functions similarly to human memory. An artificial cognitive memory can be designed to complement a computer's conventional RAM, enhancing the computer's capability with a human-like memory.

The design of the artificial memory gives insight into the workings of nature's cognitive memory. We model human memory by designing an artificial memory with human-like capabilities.

Living cognitive memory and artificial cognitive memory have features in common. All memory systems have means of information storage and means of information retrieval. Both natural and artificial cognitive memories can continue to function even if portions of their circuitry become defective. They both can have enormous capacity for information storage.

How does human memory work? How does human pattern recognition work? It turns out that these two questions are really a single question. The brain's mechanisms which achieve both are enormously complex. The purpose of this book is to find an answer, albeit a highly simplified one. The motivation is twofold: (1) to add to knowledge in the field of neuroscience and (2) to suggest the design of a highly simplified but useful cognitive memory for computers that can be either constructed from software, constructed from existing electronic hardware components, or constructed from a hybrid of software and hardware. A guiding light comes from experience of actually living with a cognitive memory.

In this book, the term *cognitive memory* could refer to both human and animal memory. Or, it could refer to an artificial, human-like memory that augments a computer's memory. The context makes it clear which it is: human and animal memory or machine memory. Cognitive memory of humans and animals is essential for thinking and decision-making. Cognitive memory for machines enhances the performance of artificial intelligence algorithms.

A cognitive memory stores sensory data and images wherever space is available, not in pre-specified locations. This is an example of content-addressable memory. Streaming input signals come from perceptual sources: eyes, ears, etc. If these input signals are *interesting*, they are stored in long-term memory. Retrieval is prompted when a connection is made between current input data and data previously stored in memory. The connection or association is performed by trainable neural networks. Training is unsupervised and, in fact, the whole memory system is unsupervised. The memory system learns from its own input data.

Pattern recognition occurs when an unknown input pattern is found to associate with patterns stored in a memory folder located somewhere in the memory. When there is a *hit*, the contents of the hit folder are sent to the memory output line. Fundamentally, pattern recognition involves making an association between an unknown input pattern and a pattern or patterns stored in memory.

With today's technology, trainable neural networks with supervised learning are being used with great success. Most applications require thousands of identified training patterns from many different classes. Once trained, the neural network is called on to identify the class of an unknown input pattern. It must make a correct choice from perhaps hundreds or thousands of classes. The information extracted from the training patterns is stored in the weights of the neural network. Once training is complete, the training patterns may be discarded.

In contrast, the cognitive memory does not discard the training patterns. The training patterns are permanently stored in memory "folders," thus comprising a database. The neural networks of the cognitive memory have much simpler tasks. Instead of one neural network, there are many neural networks, each associated with its own memory folder. These neural networks need not choose the correct class of an unknown input pattern. They simply classify the input pattern as having been seen before or not. Thus, a single binary choice. *Déjà vu*: yes or no? If it is a yes, this is a *hit*, and the input pattern is classified by association with previously stored patterns. As with all neural networks, the association can be approximate, not necessarily a perfect match.

DNA as nature's memory storage medium is a controversial concept. Most neuroscientists accept the idea that memory is stored in the brain's neural networks in the form of engrams or memory traces. In contrast, we believe that memory is not stored in neural networks, but in the brain's cell nucleus.

We believe that the nucleus is where memory is stored. We believe that the DNA in the cell's nucleus is the element of the nucleus that is nature's memory storage medium. Reasons are given for this conjecture. We believe that the purpose of the brain's neural networks is retrieval and not storage.

This book is divided into five parts:

Part I is introductory, describing human memory in everyday life.

Part II describes a highly simplified model of human cognitive memory and describes its elements.

Part III outlines various engineering applications of cognitive memory to practical problems such as face recognition, Chinese character recognition, and navigation from aerial photographs.

Part IV takes up the human side of cognitive memory as it relates to seeing, hearing, walking, and speaking. Feature detection examples are explored including Hubel and Wiesel's cat and Julesz's random dot patterns. A hypothesis for a new, long-term memory model is proposed centered upon the human brain's DNA as the primary medium for long-term storage. Within this proposed model, the human brain's activities of neurons and synapses are for memory learning and retrieval—not for long-term storage.

Part V considers the meaning of meaning. How is meaning extracted from sensory inputs and from stored data? The mechanism of thinking is explored. What is thinking? How did it start long ago? Do dogs think? Absolutely! Do mosquitoes think at some level? Absolutely! What is the role of cognitive memory in the thinking process? Having only sensory inputs and a cognitive memory, could primitive thinking have begun? Could primitive thinking have evolved to modern thinking? Is cognitive memory with sensory inputs all that is needed for thinking? Fundamental to extraction of meaning and to the thinking process is the ability to express everything derived from one's environment as images that could be recorded in a cognitive memory. People can recall the extracted images, not fine details. The meaning of things is what is remembered.

Stanford, USA

Bernard Widrow
Edward P. Katz

Acknowledgements

The authors would like to thank

Dr. Henry Yin and Dr. Gregory Kovacs who provided valuable comments and feedback on the manuscript.
Dustin Mathia transcribed Chaps. 9–12 from handwritten notes.
Yifan (James) Xu transcribed Chaps. 13–15 from handwritten notes, formatted the equations in Chaps. 5 and 6, and drew figures in Chaps. 7, 8, 13, and 21 from hand-drawn figures.

Permission was granted by Springer for the use of material from Chap. 3 of the book *Cybernetics 2.0: A General Theory of Adaptivity and Homeostasis in the Brain and in the Body*, Bernard Widrow, *Springer*, 2023.

Contents

Part I
The Cognitive Memory

Part I is an introduction to the fundamental concepts of cognitive memory. Human cognitive memory is highly complex. But some of its salient features can be modeled. These features should be apparent (though perhaps not completely recognized) to all of us who have been living with our own human cognitive memory for many years. An artificial cognitive memory can be built as a model using current software and electronic hardware technology. The model can be used to study the behavior of cognitive memory. The model can also be incorporated into a computer architecture, giving the computer a conventional memory plus a simplified human-like memory, with engineering applications.

Chapter 1
Overview

Abstract Human memory, like all memory systems, has both a means of information storage and, its counterpart, a means of information retrieval. Human memory is content addressable. An input-signal prompt begins a search of the entire memory system. Such a prompt might be an incoming visual image of an unknown object. Searching involves looking for a memory folder or folders containing information related to that prompt. If the search does make a hit in a particular folder, that folder's contents are then delivered to the memory output as retrieved information. A folder's contents might typically be sensory images such as visual, auditory, tactile, vestibular, olfactory, etc. all recorded simultaneously. Information is stored long-term at a molecular level, and is not stored in the memory's neural networks. The neural networks' function is to facilitate retrieval, not storage. A cognitive memory performs pattern recognition as well as data storage. Presence of an unknown object prompts the memory to connect the unknown object to previously encountered objects that have been identified. As we are all living with a cognitive memory, learning how to model that memory's functions can suggest how to design an artificial cognitive memory for computers that is a separate memory, complementary to a conventional computer memory.

It was another bright, cloudless, blue-sky, sunny day like so many days in Silicon Valley that found Bernie Widrow taking a leisurely stroll down the sidewalk along Palo Alto's University Avenue. A lot of activity caught his eye. People going in and out of shops, people getting in and out of parked vehicles along the sides of the street, vehicle traffic passing by in both directions, some people walking past him, other people walking on the parallel sidewalk on the other side of the street, and mothers pushing their children in strollers and baby carriages. Indeed, it was a perfect day in Palo Alto. Amidst all the people's activity, a familiar face appeared about 50 m ahead on the same sidewalk, and that person was coming directly towards Bernie. Both stopped when the diminishing distance between them was comfortable enough to engage a conversation. A very nice conversation emerged with both of them cordially participating. And all the while Bernie kept saying to himself,

> What is his name? What is his name?...What the hell is his name?

B. Widrow and E. P. Katz, *Cognitive Memory*, Springer Series on Bio- and Neurosystems 17, https://doi.org/10.1007/978-3-031-80939-2_1

Bernie and the other person each had a relaxed stance, the body language of both suggesting a friendly, casual conversation. Bernie's head gently bobbed around during the conversation, and the other person's head was also moving in a similar fashion suggesting this to be a comfortable, animated exchange between acquaintances. They were each seeing the other in somewhat different lighting, different perspectives, different distances from each other, and so forth. All the while, Bernie was listening to the other person's voice unconsciously observing their word usage and pronunciations. Some ten minutes or so into the spirited conversation, it suddenly became clear to Bernie that this fellow's name is Jonathan Jones. At the next conversation pause, Bernie injected, "*Well Jonathan, how is your wife, Mary, doing?*" Jonathan had assumed that all the while Bernie had remembered him including remembering his name.

We all have had experiences like this one. What does it teach us about how human memory works? What are the lessons to be learned from these experiences?

A few weeks later, Bernie and his wife were sitting at their kitchen table. Their conversation shifted to one of their old friends whom neither had seen for a long time. They could remember all the details about their friend, but neither could remember the friend's name. Bernie's wife is always better at remembering people's names with ease. But this time the friend's name proved elusive to both of them. They certainly did know this friend quite well, but the elusive name proved exasperating. The following day while Bernie was at his computer reading and responding to Email, the elusive friend's name which Bernie and his wife couldn't remember suddenly popped into his head. Immediately, Bernie jumped up and quickly went over to his wife and blurted out the friend's first name. "*Of course!*," she said. "*How silly*?" And then she blurted out the person's last name immediately.

1.1 Questions

The pertinent questions are:

What's going on here?

Why do these things happen?

Think about it. What do these experiences tell us about human memory? Clearly, human memory is not at all like computer memory. How the human brain functions is not at all like a conventional computer. We can remember things. But, sometimes it's hard to retrieve these things when we need or desire them.

1.2 Content Addressable Memory

In a modern computer, items of information or data are stored in specifically numbered memory locations under the control of a stored program. The program is constructed by one or more people. When data is needed, the program instructs the memory to fetch and deliver the contents from a specifically numbered location (sometimes referred to as a memory address). The program knows where to look in its main memory since it previously instructed the memory to store the item of data in that specifically numbered location.

It is safe to say that the human brain does not have numbered memory locations. Indeed, the human brain knows nothing about numbers. (A newborn baby can remember things, but has no concept of numbers.) A computer is built to process numbers, e.g., to perform arithmetic operations, manipulate "symbols" of information in its memory, etc. There are electronic circuits in the computer's central processing unit (CPU) that are designed to perform addition, subtraction, multiplication, division, and many other operations on numerical values.

There is no evidence that such circuits like this exist in the human brain. The brain is not capable of computation directly, nor is it capable of playing basketball. The brain is a general-purpose learning device which can be trained to perform these things. Brain structural composition consists largely of neurons, axons, synapses, dendrites, and much more. Nothing in the brain is innately capable of performing computation or any form of mathematics. Mathematics is a human artifact.

Information and data are stored in the brain where there are empty places. The human user of this information has no idea where specific data is stored. The brain has no stored program to direct the memory locations for either storage or retrieval of specific data. There are no numbered storage locations.

However, there is evidence that our memory is stored chronologically in time and sequentially over spatial location. In some cases of head injury, people have lost memory over a span of years. Early memory may not be lost. The person is able to speak and walk. The ability to record and retrieve new memory since recovering from the injury remains intact. This suggests that a local portion of the brain storing data for a given epoch of time has been destroyed as a result of the injury.

1.3 Information Retrieval from Content Address Memory

Once information or data is stored in human memory, there is no way for the human user to know where this is stored. Data is retrieved in response to prompt inputs. Signals that are sent to the entire memory system cause the memory to deliver the desired data, wherever its location. From the previous Jonathan Jones example, the image of his face and then the sound of his voice were prompts that yielded the knowledge to enable a cordial conversation and ultimately his name retrieval. For the second example above, the initial discussion between Bernie and his wife about the familiar mutual friend was not an adequate prompt to produce the person's name. However, the next day the first name popped into Bernie's head. Saying the friend's first name to his wife was a sufficient additional prompt to allow her to recall the person's entire name and many more remembered things about that person. All of this memory was stored in their heads–the issue that they had was with retrieval. Storing things is relatively easy. Recalling them is generally more difficult and is sometimes problematic. Nevertheless, human memory is amazing and worth studying by engineers who would like to design and build a human-like memory for computers. Studying human memory is fundamental for neurobiology and psychology. Many disciplines are involved in the quest to understand human memory.

1.4 Pattern Recognition

Pattern recognition is an activity which engages us all day long. We are constantly observing things, checking them, looking for the familiar, looking for something interesting, looking for something tasty, looking for something provocative, and looking out for potential danger. Fundamentally, pattern recognition involves first sensing the real-world environment by seeing, hearing, feeling, smelling, etc. And second, it involves trying to relate what is sensed to previous experiences stored in memory. An example might be the following:

> A small child seeing a wooden toy block on a table first recognizes it and then plays with it. The child knows all about the block, having played with it the day before, when the block was picked up and dropped, felt and gripped, viewed from many angles, chewed on, etc.

Visual images of the block, the sound of the block being dropped onto the table (or floor), the tactile feel of the block, even the taste of the block, are all images or patterns stored in the child's memory. Seeing the block a second time provides a visual image that acts as a prompt and causes knowledge about the block to be recalled. The child is immediately familiar with the block and wants to play with it again, maybe to learn more about it.

Pattern recognition certainly depends on the brain's memory, and learning involves storing new information and data into the memory and retrieving previously learned memory. The question now is,

> How does all this work?

Most scientists in the field of neurobiology believe that memory is stored within the vast network of neurons that comprise certain portions of the brain. These are networked together communicating with each other by signaling electrically and chemically. Connections between the neurons are made through synapses. These synapses vary in their ability to conduct and amplify neural signals. Animal studies have shown that synaptic conduction varies with experience. From this emerges the belief in the idea that memory is stored in the neural network synapses. Further support for this belief comes, for example, from the observation that a disease like

Alzheimer's attacks and destroys the neurons and synapses which causes memory loss.

1.5 Alternate View of the Brain's Memory Storage

We have good reasons to believe that the idea that human memory, especially long-term memory, being stored in the neural network synapses is incorrect. Rather, that memory is stored at the molecular level. The neural network's role in the brain's memory function is primarily for retrieving information, not for the actual storage of the information itself. More will be said about this later.

Bernard Widrow's research into artificial neural networks began in 1957 and has continued probably longer than any other continuous research effort into memory systems. Artificial neural networks are certainly not the same as the real ones in the brain. Yet, they both have similar behavior characteristics. As an Electrical Engineering graduate student at MIT in 1951, Bernie joined the group responsible for inventing and developing the magnetic core computer memory, a random-access memory that was a forerunner of modern computers' main memory of RAM chips. From this experience, Bernie learned about computer memory and about computer hardware construction and function. Today's computers and their memories are structurally quite similar to those emerging in the 1950s. Prof. Widrow's knowledge of computer memory and artificial neural networks gives him an interesting background of experience for the study of human memory since he has lived with his human memory for a very long time. He speaks to other people who have lived with their human memories. He asks himself as a memory designer, "How could I build an artificial memory that would have similar operational characteristics like those of human memory?" In designing such a "*cognitive*" memory, he expects useful advances to result in computer memory and in the field of computation. Hopefully this will give neurobiologists new ideas for advancing their research and brain modeling studies. In some sense, this book is both a progress report and a call to action.

1.6 Summary

This chapter introduces the idea of cognitive memory and provides some explanation of its workings. Understanding is focused on human memory. Examples of memory performances (or lack thereof) are presented. They provide some insight which enables the modeling of human memory. And they also can suggest the design of artificial cognitive memory for potentially augmenting conventional computer memory. A cognitive memory both stores information and performs pattern recognition and information retrieval.

Appendix

Use google (or your favorite internet search engine) to explore:

- Content addressable memory
- Alzheimer's disease
- Brain injury
- Remembering names
- Phineas Gage

Chapter 2
Cognitive Memory

Abstract A cognitive memory does the following:

- Stores sensory input patterns.
- Stores patterns where space is available.
- Stores visual, auditory, etc. sensory input patterns simultaneously in the same folder.
- Data recovery is triggered by prompt input patterns.
- Autoassociative neural networks are used in data retrieval.
- The visual cortex, auditory cortex, etc., are used for pattern pre-processing—not for pattern storage.

Cognitive memory can be a machine memory which can augment conventional computer memories. It's intended to function like human memory, as we currently understand the brain's memory functionality.

2.1 Development of a Cognitive Memory

When embarking on a path towards developing a cognitive memory, three important questions arise:

1. How does human memory work?
2. How could such a memory be built?
3. How could such a memory be used to solve practical problems?

What would we like to do? The main intent is to design a memory system that is as simple as possible, but this has to behave like and to emulate human memory. Why would we like to do this? Developing this new kind of memory for computers that works as an adjunct to existing forms of computer memory, and facilitates solutions to problems in artificial intelligence, pattern recognition, speech recognition, control

B. Widrow and E. P. Katz, *Cognitive Memory*, Springer Series on Bio- and Neurosystems 17, https://doi.org/10.1007/978-3-031-80939-2_2

systems, etc. Another reason to do this is to advance cognitive science with new insights into the workings of human memory.

2.2 Features of a Cognitive Memory

The following are salient features of cognitive memory:

- Stores sensory patterns (visual, auditory, tactile, radar, sonar, medical images, etc.), as in a conventional computer memory.
- Stores patterns wherever space is available, not in specifically numbered memory locations.
- Stores simultaneously sensed input patterns into the same *folder* (e.g., simultaneous visual and auditory patterns are stored independently but together). The idea for a cognitive memory folder is like a computer's file folders; a folder is a metaphor to describe a function needed for human memory – there are no actual folders in the human brain.
- Data retrieval is in response to "prompt" input patterns (e.g., either a visual or auditory input pattern, or both can trigger recall).
- Autoassociative neural networks are used in the data retrieval process (artificial neural networks and autoassociative versions of them will be described later).

The reasoning behind these features comes from experience with one's own memory. All of these features can be realized with present day software and electronic technology. The idea that information and data are stored in the brain's memory seems obvious. The idea that all sensory data observed, sensed, and being recorded at the same time are stored in the same folder is not so obvious, until one thinks about it.

2.3 Prompting with Sensory Inputs

You can hear a singer's voice, and in your mind's eye you visualize that singer's face. Or, you see a photograph of the singer, and in your mind you can hear them singing a favorite song. Hearing a song prompts one to recall a vision of the singer's face. This would be difficult to accomplish if the sound and image were not stored together, as is done when a video is recorded.

2.4 Visual and Auditory Cortexes

Memory of the song and the face are not stored in either the auditory or visual cortex. The auditory cortex is located in both of the brain's hemispheres, just inside the locations of the ears. The brain's visual cortex is located just inside the back of the head. Memory of sensory input patterns (video, audio, tactile, vestibular, etc.) must be stored together somewhere else. Exactly where is a controversial issue. Now, suppose you go into a restaurant, and you are delighted with a familiar smell. In your "mind's eye," you perceive an image of your mother in her kitchen cooking your favorite meal with that same familiar set of smells.

2.5 Neural Networks for Data Retrieval

Incidentally, if the visual and auditory cortexes do not store visual or auditory patterns or data, then what exactly are their functions? This is a very good question that will be discussed further.

The idea that neural networks are used for the memory retrieval process and not for memory storage is itself not obvious. This is another important subject that will be discussed in detail subsequently.

2.6 Summary

Input data or information is stored wherever there is empty storage space. Once stored, information is retrieved in response to a prompt. Autoassociative neural networks can be used to establish a connection or correlation between the prompt and a folder containing relevant information. Modeling human cognitive memory gives insight to the design of artificial cognitive memories for computers.

Appendix

Use google (or your favorite internet search engine) to explore:

- Prompting for memory retrieval
- Autoassociative neural networks
- Pre-processing by visual cortex, auditory cortex

Part II
Autoassociative Neural Networks and Cognitive Memory Design

It is possible to build, in hardware or software (or both), a simple memory system that would "resemble" human memory. The complexity of human memory should not be underestimated. By resemble, we mean limited but still useful. We propose a piece of engineering based on experiences with one's own memory, that is intended for application to data storage and pattern recognition problems. Where there is resemblance in behavior to the behavior in human memory, insight will be gained for both engineering and biology.

Neural Networks, LMS, and Backpropagation will all play major roles in the development of a cognitive memory system. A brief introduction to neural networks will be given. Almost all neural networks are trained today by means of the backpropagation algorithm. This is an algorithm for layered neural networks that utilizes stochastic gradients and the method of steepest descent to effect adaptation and learning. Backpropagation is a *least squares algorithm* that is based on the much simpler Least Mean Squares (*LMS*) algorithm that itself is used worldwide in digital telecom systems for equalization and echo canceling and in adaptive filtering applications to noise canceling, adaptive antennas, etc.

Chapter 3 will explain the LMS algorithm, and Chap. 4 will explain the backpropagation algorithm derived from LMS. Chapter 5 will introduce autoassociative neural networks that play a major role in the design of cognitive memories.

Chapter 3
The LMS Algorithm

Abstract The adaptive linear combiner is the fundamental building block of all neural networks, and of all adaptive filters in signal processing and control applications. The coefficients or weights of the linear combiner are adjusted automatically by an adaptive algorithm. In most cases, this algorithm is known as LMS (Least Mean Squares). The linear combiner is trainable. The training involves input vectors or patterns presented to the combiner, along with corresponding desired responses. The training objective is to adjust the combiner's parameters so as to cause the pattern responses to match to the greatest degree as possible the desired responses (in the least mean squares sense). The mean-square-error is a quadratic function of the parameters. A stochastic gradient is used with the method of steepest descent to seek the optimal solution at the bottom of the quadratic "bowl." The resulting LMS algorithm is perhaps the simplest of all adaptive algorithms. It is used widely in engineering applications and appears to be a natural algorithm, part of nature's process for learning and adaptation. This chapter is based verbatum on Chapter 3 of Cybernetics 2.0 [8].

The fields of artificial neural networks (ANN's) and neuroscience have progressed over the years more or less independently. Although the ideas and motivation for ANN's originated from early knowledge of neurobiology, modern ANN's do not function like the biological neural networks found in the brain. Connections between the two fields have been proposed.

This book follows that tradition. An unsupervised learning algorithm, the Hebbian-LMS algorithm has been devised to have engineering applications, but it will very likely be a good model for natural synaptic plasticity. This algorithm fits the Hebbian learning rules. It is an implementation of Hebbian learning with the LMS algorithm. The LMS (least-mean-squares) algorithm is well known in Computer Science and Electrical Engineering.

B. Widrow and E. P. Katz, *Cognitive Memory*, Springer Series on Bio- and Neurosystems 17, https://doi.org/10.1007/978-3-031-80939-2_3

3.1 Contribution of Norbert Wiener

Least squares methods were probably invented by Gauss, or possibly by Leibniz before him. Modern statistical methods of regression are based on this. An example would be fitting a parabolic curve to a set of data points. The parameters of the parabola would be adjusted so that the errors of each data point would be minimized. A "best fit" would minimize the sum of squares of the errors.

Why minimize mean-square-error? It makes sense. Positive errors are weighted equivalently to negative errors. Why not minimize the magnitude of errors, or minimize the mean of the magnitude of the errors cubed or minimize the mean fourth power of the errors? The answer is that they all make sense but minimizing the mean-square-error leads to simple elegant analytic solutions to problems.

Norbert Wiener chose the least square criterion and brought it into the world of engineering. He was the first to design statistically optimal signal processing filters for signal prediction and noise reduction. Nothing like this had ever been done before. It was revolutionary. In Wiener's day, filters were constructed using analog circuits. These were usually designed to either be low pass, band pass, or high pass filters, but there were no filtering methods used based on a statistical foundation. Wiener's method requires the prior knowledge of the autocorrelation of the input signal and the cross correlation between the input and the desired response. Wiener introduced the concept of the desired response. The method of Widrow et al., did not require knowledge of the signal statistics, but adapted based on input data.

Following in Wiener's footsteps, Widrow and his students at MIT in 1957 began simulating adaptive filters and adaptive neurons that learned from presented examples. The learning capability was based on the gradient descent approach. These were all digitally simulated using an IBM mainframe.

In 1959, at Stanford, Widrow and his first Ph.D student Marcian (Ted) Hoff, Jr. invented or discovered the LMS algorithm. If they invented it, then they were doing engineering and applied mathematics. If they discovered it, then nature invented it eons ago, and they were doing physics.

3.2 The Linear Combiner and the LMS Algorithm

In the discussion to follow, it should be noted that signals in artificial neural networks are not represented by the frequency of pulse trains as in living neural networks, but by variable (dc) levels. Learning in artificial neural networks is not done continuously, but cyclically at discrete time intervals. Otherwise, the analogy between biological and artificial versions of neurons, synapses, and networks will be apparent.

In nature's neural networks, the input signals to each neuron are summed *electrochemically*. In artificial networks, the neuronal input signals are summed numerically. We begin with the LMS (least-mean squares) algorithm for adapting the linear combiner's weights.

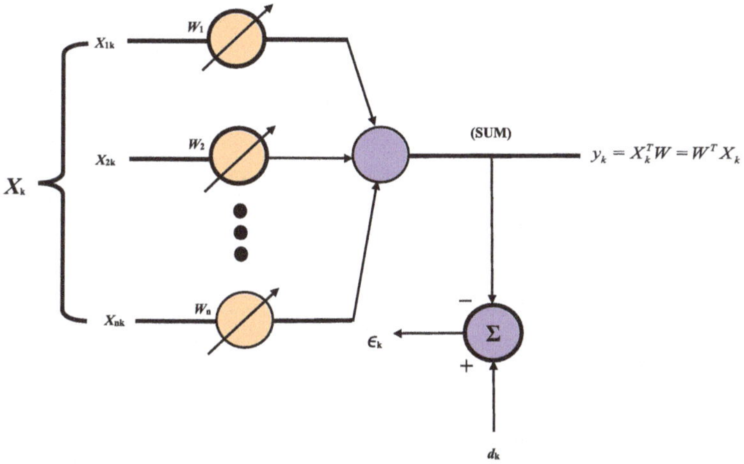

Fig. 3.1 A trainable linear combiner

A basic building block of all neural networks, biological or artificial, and of adaptive signal processing and control systems, is the trainable linear combiner diagrammed in Fig. 3.1. During training, input vectors or patterns are presented to the input terminals. Each component of each input vector is weighted, and the weighted sum is computed to provide an output. The kth input pattern or input vector is designated as X_k. The output for that input is the (SUM), $y_{k.}$ The weights are the components of the weight vector W. Thus,

$$y_k = X_k^T W = W^T X_k \tag{3.1}$$

During training, each input vector, X_k, is supplied with a corresponding desired response d_k. The purpose of training is to cause the linear combiner to produce outputs for the entire set of training patterns that most closely matches the respective desired responses, in the least-squares sense.

For each input vector, the error ϵ_k is the difference between the desired response d_k and the actual output y_k.

$$\varepsilon_k = d_{k-}y_k = d_k - X_k^T W \tag{3.2}$$

The square of the error is

$$\varepsilon_k^2 = d_k^2 - 2d_k X_k^T W + W^T X_k X_k^T W. \tag{3.3}$$

The mean-square-error (MSE) is the mean value or expected value of (3.3)

$$\text{MSE} = E\left[\varepsilon_k^2\right] = E[d_k^2] - 2E[d_k X_k^T]W + W^T E\left[X_k X_k^T\right]W. \tag{3.4}$$

The cross-correlation between the desired response and the input vector can be represented by the vector P.

$$E[d_k X_k^T] \triangleq P \tag{3.5}$$

The input autocorrelation matrix can be represented by the matrix R

$$E[X_k X_k^T] \triangleq R \tag{3.6}$$

If the input vectors have zero bias, the matrix R is called the *input covariance* matrix.

Substituting (3.5) and (3.6) in (3.4), we have

$$\text{MSE} = E[\varepsilon_k^2] = E[d_k^2] - 2PW + W^T RW \tag{3.7}$$

Thus, the mean-square-error is a quadratic function of the weights. A sketch of a typical MSE is shown in Fig. 3.2. The MSE surface is a paraboloid. With many weights, this surface becomes a hyperparaboloid.

During adaptation, the weights are varied to minimize the mean-square-error. One way of doing this makes use of the method of steepest descent. A series of steps is taken starting from the initial guess for the initial weight vector and following the negative gradient, step by step approaching the bottom of the quadratic bowl. This is like a ball rolling down a hill.

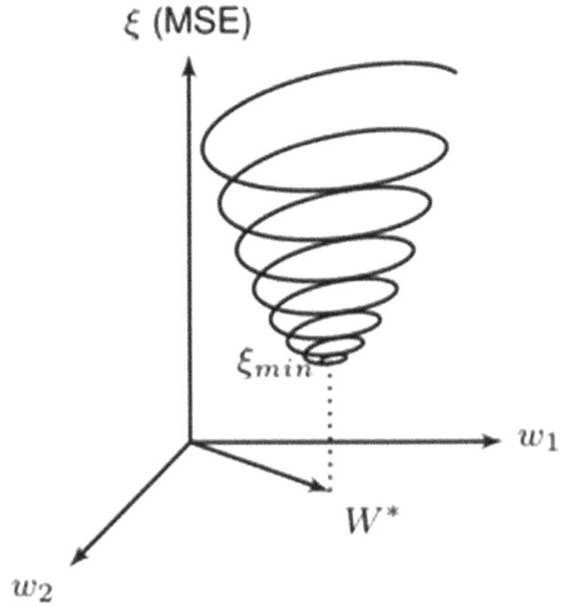

Fig. 3.2 : A mean-square-error surface. A paraboloid

Accordingly

$$W_{k+1} = W_k + \mu(-\nabla_k) \tag{3.8}$$

The next weight vector equals the present weight vector plus a constant μ multiplied by the negative gradient. The parameter μ determines step size, stability, and speed of convergence.

The gradient at the kth iteration is defined as

$$\nabla_k \triangleq \frac{\partial E[\varepsilon_k^2]}{\partial w_k} = \begin{bmatrix} \frac{\partial E[\varepsilon_k^2]}{\partial w_{1k}} \\ \vdots \\ \frac{\partial E[\varepsilon_k^2]}{\partial w_{nk}} \end{bmatrix} \tag{3.9}$$

The components of the gradient vector are the partial derivative of MSE with respect to the individual weights. The number of weights is n.

Differentiating Eq. (3.7) yields the gradient,

$$\nabla_k = -2P + 2RW_k. \tag{3.10}$$

The optimal solution $W*$ is obtained by setting the gradient ∇_k to zero. Accordingly,

$$-2P + 2RW^* = 0 \tag{3.11}$$

The optimal weight vector $W*$ is

$$W^* = R^{-1}P \tag{3.12}$$

and this is called the *Wiener solution.*

The Wiener solution (3.12) often cannot be used in practice because one only has input data and would not have the correlation functions, P and R. In practice, the Wiener solution is an idealization, and its performance is a benchmark.

Minimizing the MSE is generally done with the method of steepest descent.

$$W_{k+1} = W_k - \mu\nabla_k \tag{3.13}$$

Following the negative gradient is like the action of a ball rolling down a hill.

An analytic expression for the gradient would not be available; only input data would be available. The idea of Widrow and Hoff was to take the square of a single sample of the error associated with a single input vector and call that the MSE. Of course, this is not the case being that it is only a single error sample. Differentiating, it yields a crude "noisy" gradient estimate,

$$\hat{\nabla}_k = \frac{\partial \varepsilon_k^2}{\partial w_k} = 2\varepsilon_k \frac{\partial \varepsilon_k}{\partial w_k}. \tag{3.14}$$

The expression for the error ε_k is given by Eq. (3.2). Differentiating, one obtains the crude gradient estimate as

$$\hat{\nabla}_k = 2\varepsilon_k \frac{\partial \varepsilon_k}{\partial w_k} = 2\varepsilon_k(-X_k) \tag{3.15}$$

The expected value of this gradient estimate can be proven to be the true gradient. The crude estimate is unbiased. Using the gradient estimate with steepest descent,

$$W_{k+1} = W_k - \mu \hat{\nabla}_k \tag{3.16}$$

From this we have the LMS algorithm:

$$W_{k+1} = W_k + 2\mu\varepsilon_k X_k \tag{3.17}$$

$$\varepsilon_k = d_k - x_k^T W_k \tag{3.18}$$

The LMS algorithm can be described in words: "*With the presentation of each input pattern vector and its associated desired response, the weight vector is changed slightly by adding the pattern vector to the weight vector, making the sum more positive, or subtracting the pattern vector from the weight vector, making the sum more negative, changing the sum in proportion to the error in a direction to make the error smaller*". The parameter μ is generally chosen to be small. Many small steps on average follow the gradient toward the bottom of the bowl. In the limit, as the number of iterations increases, the expected value of the weight vector exponentially approaches the Wiener solution (3.12). The properties of the LMS algorithm are discussed in detail in the text of Widrow and Stearns [1], and in many other sources [2]. The original 1959 Widrow-Hoff paper is reference [3].

The LMS algorithm is used in adaptive filters and artificial neural networks. This algorithm is used in one form or another for channel equalization and echo canceling in telecommunication systems. Today, pervasive uses of this can be found in every MODEM in the world as well as being one of the Internet's enabling technologies. It is the foundation of modern day neural networks based on the backpropagation algorithm, a gradient descent algorithm. The crude gradient estimate was a "crazy idea," but it turned out to be a pretty good crazy idea.

3.3 Steepest Descent, a Feedback Algorithm

The method of steepest descent is a feedback algorithm. A block diagram of the method of steepest descent applied to minimization of mean-square-error when MSE is a quadratic function of the weights is shown in Fig. 3.3.

To verify that this feedback system does indeed represent the method of steepest descent, begin the verification of the point marked, W_{k+1}. Delaying W_{k+1} by one cycle time gives W_k.

From the diagram, W_{k+1} is

$$W_{k+1} = W_k + \mu(-\nabla_k) \tag{3.19}$$

This agrees with Eq. (3.13) and does represent steepest descent. Referring again to the diagram, the gradient is seen to be

$$\nabla_k = 2R\big(W_k - W^*\big) \tag{3.20}$$

Substituting the expression for the Wiener solution,

$$\nabla_k = 2R\big(W_k - R^{-1}P\big) = -2P + 2RW_k \tag{3.21}$$

This confirms Eq. (3.10).

Referring again to the diagram, when the system converges, $W_{k+1} = W_k$ and accordingly the gradient $\nabla_k = 0$. Therefore, $W_k = W^*$. The weight vector W_k has converged on the Wiener solution W^*.

When employing the method of steepest descent, one does not implement the feedback system per se. One simply follows the gradient, starting from an initial condition, step by step toward the bottom of the quadratic bowl, like a ball rolling

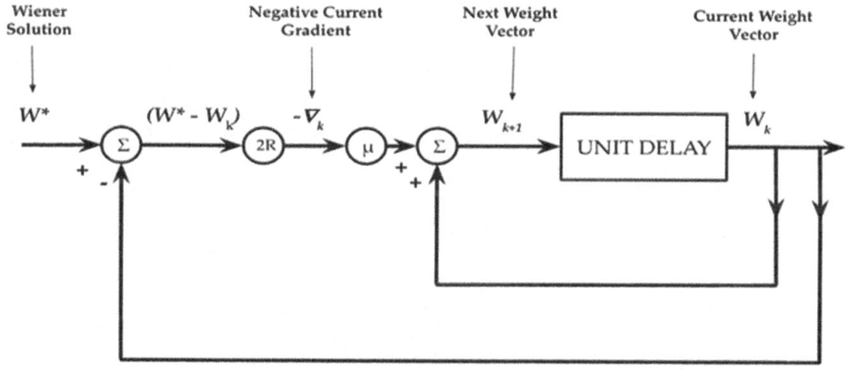

Fig. 3.3 Feedback diagram for a quadratic MSE surface, controlling the weight vector

down a hill. The initial value of W_k is set to be a best initial guess of W^* which the iterative process of steepest descent will ultimately find.

The feedback system is a model of what happens and gives insight into the process. The "error" signal is the difference ($W^* - W_k$). Negative feedback is used to minimize this error. The feedback system is linear with a transient response that is a sum of exponentials. The number of exponential components is equal to the number of distinct eigenvalues of the input autocorrelation matrix R. If all of the eigenvalues are equal, the quadratic mean-square-error surface is circular. An expression for the single time constant under this condition can be shown to be

$$\tau = \frac{n}{4\mu \text{trace} R} \tag{3.22}$$

The number of weights is n. Time τ is measured in the number of adapt cycles.

The LMS algorithm does not perfectly perform the method of steepest descent since the LMS gradient is not the true gradient because it is an estimate. The feedback diagram can be used to represent this by adding noise to the gradient. The noise propagates in the feedback system and the result is noise in the weight vector W_k. At convergence, the weight vector does not go to the Wiener solution but varies randomly about it. This creates an excess mean-square-error. The average excess mean-square-error divided by the minimum mean-square-error of the Wiener solution is a dimensionless measure of how far on average the LMS solution is from the ideal Wiener solution. This ratio is called the *misadjustment.* A formula for misadjustment of the LMS algorithm is

$$M = \mu(\text{trace} R) \tag{3.23}$$

All adaptive systems that learn from input data must have misadjustment. This is because the number of learning samples is finite, not infinite, that would be needed for perfect steepest descent.

It is important to note that increasing the size of μ makes convergence faster (inversely related to learning time constant) but increases the misadjustment. In engineering practice, there is a tradeoff between speed of convergence and misadjustment. A good starting point would be to choose μ such that misadjustment would be approximately 10%. This would result in a learning time, four times the learning time constant, to be a number of training cycles equal to ten times the number of weights for a linear combiner or for a single artificial neuron. Combining Eqs. (3.22) and (3.23), a relation between learning time and misadjustment is obtained:

$$4\tau = \frac{n}{M} \tag{3.24}$$

The learning time, 4τ, equals $\frac{n}{M}$.

For a given level of misadjustment, the learning time increases in proportion to the number of weights. A complete discussion of this subject is given in Widrow and Stearns [1] and Widrow and Wallach [4].

The simplicity of the LMS algorithm is unrivaled. On the face of it, who would expect such a simple algorithm to be using stochastic gradient descent to minimize mean-square-error, a quadratic function of the weights? Who would expect such a simple algorithm to play a major role in learning in the biological brain?

The LMS algorithm has been used worldwide for many years. Its performance is "bullet proof." This reflects favorably on the solid stable behavior of living neural systems.

Applications of the LMS algorithm can be found in the references [5–8].

3.4 Summary

The LMS (Least Mean Square) algorithm is intended for the adaptation of the coefficients of a linear combiner, which is a key element of a single neuron. Training patterns (vectors) and their associated desired responses are presented and the difference between the output, the sum (SUM) and the desired response for a given training pattern is the error. The LMS algorithm, based upon the method of steepest descent, seeks an optimal solution and an optimal set of coefficients or weights that minimize the mean-square-error, averaged over the set of training patterns. The optimal solution is called the Wiener solution. The basic LMS algorithm is "supervised," in the sense that a desired response is specified for each input training pattern. An "unsupervised" version of LMS has input training patterns, but desired responses are not given. The unsupervised version is the basis for the Hebbian-LMS algorithm.

Appendix

Use google (or your favorite internet search engine) to explore:

- Adaptive filtering, LMS algorithm

References

1. Widrow, B., Stearns, S.D.: Adaptive Signal Processing. Prentice Hall, Englewood Cliffs, NJ (1985)
2. Haykin, S.S.: Adaptive Filter Theory. Pearson Education India, Noida (2008)
3. Widrow, B., Hoff, M.E., Jr.: Adaptive switching circuits. IRE WESCON Conv Record **4**, 96–104 (1960)

4. Widrow, B., Walach, E.: Adaptive Inverse Control, Prentice Hall, Upper Saddle River, NJ, 1996; Reissue Edition: Adaptive Inverse Control—A Signal Processing Approach. Wiley, Hoboken, NJ (2008)
5. Widrow, B., et al.: Adaptive noise canceling: principles and applications. Proc. IEEE **63**(12), 1692–1716 (1975)
6. Lucky, R.W.: Automatic equalization for digital communication. Bell Syst. Tech. J. **44**(4), 547–588 (1965)
7. Haykin, S., Widrow, B. (eds.): Least-mean-square adaptive filters. Wiley, Hoboken, NJ (2003)
8. Widrow, B.: Cybernetics 2.0: A General Theory of Adaptivity and Homeostasis in the Brain and in the Body. Springer, Cham (2022)

Chapter 4
Adaline

Abstract The simplest and most widely used learning algorithm is the least mean square (LMS) algorithm. This algorithm was used to train ADALINE (adaptive linear neuron). ADALINE was a single trainable neuron. Input training patterns were presented, and the weights (coefficients) were adjusted to deliver neuron outputs that are the best least square fit to the desired outputs of the training patterns. LMS is based on the method of steepest descent. ADALINE is the basic building block of present-day neural networks.

In the late 1950s and early 1960s, two trainable neural elements were in development in the United States. One was Rosenblatt's Perceptron and the other was Widrow and Hoff's ADALINE. They were similar but used different training algorithms. ADALINE was trained using the LMS algorithm [1], and that training method has been in continuous use since then. ADALINE (adaptive linear neuron) was built as a piece of hardware. A photograph from 1960 is seen in Fig. 4.1. It is a trainable, single-neuron pattern classifier. A schematic diagram of ADALINE is shown in Fig. 4.2.

The photo of ADALINE (Fig. 4.1) shows a 4×4 array of toggle switches by which patterns are inserted. These are crude 16-pixel patterns. With a switch up, the pixel input is $+1$. With a switch down, the pixel input is -1. Each toggle switch is connected to a potentiometer that implements a single weight. The potentiometers are adjusted by rotating the white knobs. The linear (SUM) is s, the sum of the products of each component of the input vector and its corresponding weight. (Mathematicians refer to this either as a *scalar product,* an *inner product* or a *dot product*). The meter indicates this (SUM). The linear error ϵ is the difference between (SUM) and the desired response, which is either $+1$ or -1.

A quantized output y is taken as the sign of the (SUM). Thus, input patterns were classified as $+1$'s or -1's, as a bipolar binary classification.

An algebraic description of this is

$$s_k = X_k^T W_k \leftarrow \text{lincar output}$$
$$\epsilon_k = d_k - s_k \leftarrow \text{linear error}$$

B. Widrow and E. P. Katz, *Cognitive Memory*, Springer Series on Bio- and Neurosystems 17, https://doi.org/10.1007/978-3-031-80939-2_4

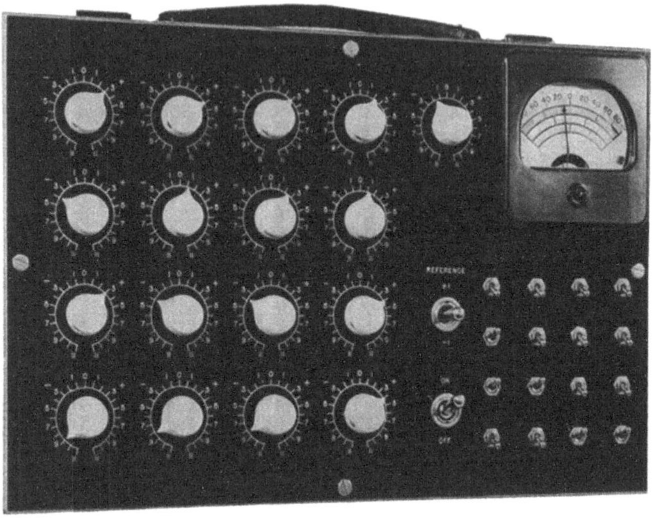

Fig. 4.1 Knobby ADALINE, built in 1960

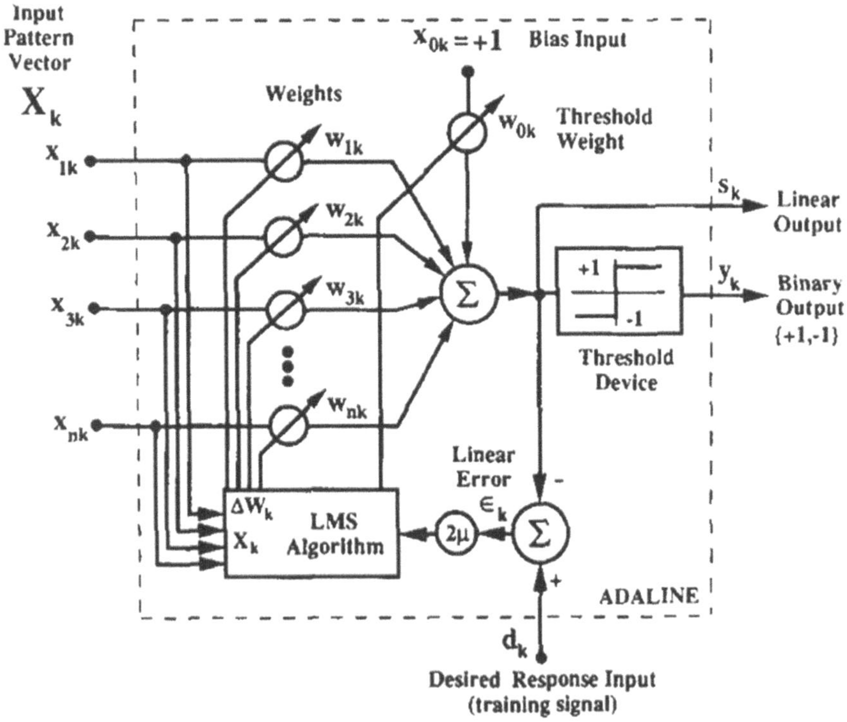

Fig. 4.2 Adaptive linear element (ADALINE)

$$y_k = \text{SGN}(X_k^T W_k) \leftarrow \text{quantized output, output decision}$$

where X_k^T is the transpose of the k_{th} input pattern and W_k is the k_{th} weight vector. Note that SGN represents the signum, i.e. the sign of the result (either + 1 or − 1).

The ADALINE seen in the photo (Fig. 4.1) is operated in the following manner. An input pattern is presented by means of the 4 × 4 array of switches. Suppose that the desired response for that pattern is +1, i.e. that pattern should be classified as a +1. Observing the meter reading, an error is noted as the difference between +1 (halfway on the right-hand scale of the meter) and the meter reading. If the reading is below +1, a correction of the weights (via the "knobby" potentiometers) is needed to bring the response closer to the desired response. In accordance with the LMS algorithm, the input pattern, which is a vector, is added to the weight vector to cause the (SUM) to be more positive. Before adding, a scale factor proportional to the error is applied to the input vector. The amount of the error corrections will be proportional to that error.

Now suppose that the input pattern is inserted, the desired response or pattern classification remains +1, but the meter reading (SUM) is more positive than +1. To reduce the error, (SUM) must be reduced. Scaled by the error, the input pattern should be subtracted from the weight vector.

It should be noted that this error correction procedure would also be the same if the desired response were +1. Namely, add the pattern vector to the weight vector to increase (SUM), or subtract the pattern vector from the weight vector to reduce (SUM) scaled by the magnitude of the error. This is done in the direction to reduce the error.

The weights are trained with each input pattern presentation to make the error smaller for that input pattern. By repetitively doing this and recycling the input patterns over and again, this error correcting algorithm will cause the weights to be adjusted to reduce the mean-square-error (MSE) over the set of input patterns. If the number of training patterns is less than or equal to the number of weights, the errors can all be brought to zero. Thus, the linear capacity of the single neuron is equal to the number weights.

To train ADALINE to make all the right decisions, the errors need not to be brought to zero. The desired classifications can be achieved without the errors being zero as a result of the quantization produced by the SGN function. This function is known as the "*activation function*" of the neuron.

Training ADALINE becomes very simple and intuitive with a little practice. A youtube video showing Prof. Widrow training ADALINE is titled "The LMS Algorithm and ADALINE. Part 1—The LMS algorithm[1]" [1]. The ADALINE of Fig. 4.1 was used in some of the earliest experiments for training classifiers. This was before computers were pervasively available in individual laboratories. Over the years, the ADALINE was used in classrooms to teach students about trainable classifiers and the LMS algorithm.

[1] https://youtu.be/hc2Zj55j1zU.

4.1 Summary

Input pattern vectors are presented to ADALINE for training. The weights of the ADALINE are manually adjusted to minimize mean-square-error (MSE). For each input training pattern, the error is the difference between the neuron output and the desired response. LMS minimizes the mean square of the error, averaged over the set of training patterns. The trainable neuron is a basic building block of modern neural networks.

Appendix

Use google (or your favorite internet search engine) to explore:

- ADALINE neuron
- LMS algorithm
- Stochastic gradient descent
- Perceptron
- Fronk Rosenblatt.

Reference

1. The LMS Algorithm and ADALINE. Part 1—The LMS algorithm, YouTube, https://youtu.be/hc2Zj55j1zU

Chapter 5
Sigmodal ADALINE

Abstract A neuron is a linear combiner followed by a non-linear activation function. The activation function of the original ADALINE was a sharp quantizer that provided binary outputs for each input training pattern. Widely used in modern artificial neural networks is the more biological activation function called the *sigmoid.* A typical sigmoid function is the arc tangent. This provides an analog output that is bounded by +1 and -1. The sigmoid function is smooth and differentiable. At least squares training algorithm for a neuron with a sigmoidal activation function is derived. A trainable sigmoidal ADALINE is a basic building block of multi-layer neural networks.

The nonlinear element of the original ADALINE was a sharp quantizer, the SGN, signum. For multilayered networks of ADALINE trained with backpropagation, a smoother differentiable, nonlinear element is required. The latest version of ADALINE has the SGN replaced with SGM, the sigmoid. A linear combiner with a sigmoidal nonlinearity is shown in Fig. 5.1.

5.1 The Sigmoid

The ADALINE elements considered thus far obtain their outputs with hard-limiting quantizers (signums). The input–output mapping of the hard-limiting quantizer is $_k = \text{sgn}(s_k)$. Other forms of nonlinearity have come into use [1], primarily of the sigmoidal type. These nonlinearities provide saturation for decision making, yet they have differentiable input–output characteristics that facilitates adaptivity. We generalize the definition of the ADALINE element to include the possible use of a sigmoid in place of the signum, and then determine suitable adaptation algorithms. Figure 5.1 shows a "sigmoid ADALINE" element which incorporates a sigmoidal nonlinearity. The input–output relations of the sigmoid can be denoted by $_y = \text{sgm}(\text{s}_\text{k})$. A typical sigmoid function is the hyperbolic tangent:

B. Widrow and E. P. Katz, *Cognitive Memory*, Springer Series on Bio- and Neurosystems 17, https://doi.org/10.1007/978-3-031-80939-2_5

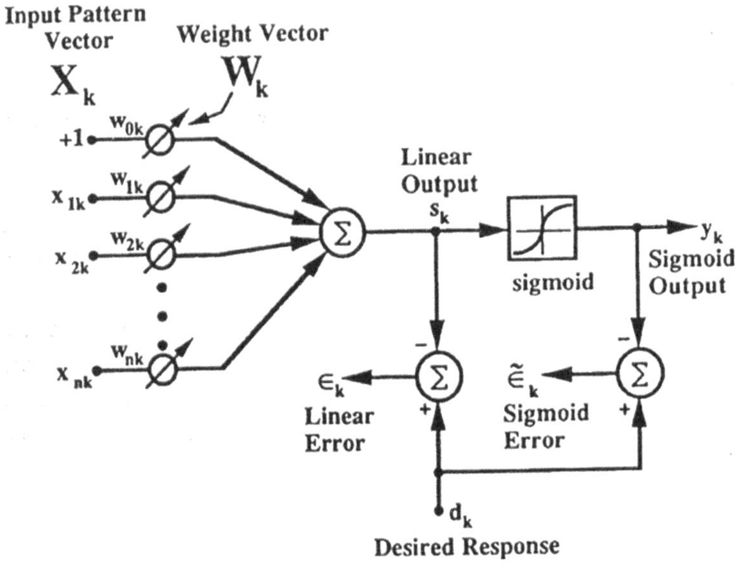

Fig. 5.1 Adaline with sigmoidal nonlinearity

$$y_k = \tanh(s_k) = \left(\frac{1 - e^{-2S_k}}{1 + e^{-2S_k}}\right) \tag{5.1}$$

We will adapt this ADALINE with the objective of minimizing the mean square of the sigmoidal error $\tilde{\varepsilon}_k$, defined as

$$\overline{\varepsilon}_k \triangleq d_k - y_k = d_k - \text{sgm}(s_k) \tag{5.2}$$

5.2 Backpropagation for the Sigmoid ADALINE

Our objective is to minimize $\mathbf{E}[(\hat{e}_k)^2]$, averaged over the set of training patterns, by proper choice of the weight vector. To accomplish this, we shall derive a backpropagation algorithm for the single sigmoid ADALINE element. An instantaneous gradient is obtained with each input vector presentation, and the method of steepest descent is used to minimize error as was done with the LMS algorithm of Eqs. (3.16), (3.17), and (3.18).

Referring to Fig. 5.1, the instantaneous gradient estimate done with the LMS algorithm of Eqs. (3.16), (3.17), and (3.18) obtained during the presentation of the kth input vector χ_k is given by

$$\widehat{\nabla}_k = \frac{\partial(\overline{\varepsilon}_k)}{\partial W_k} = 2\overline{\varepsilon}_k \frac{\partial \overline{\varepsilon}_k}{\partial W_k} \tag{5.3}$$

Differentiating Eq. (5.2) yields

$$\frac{\partial \overline{\varepsilon}_k}{\partial W_k} = \frac{\partial \mathrm{sgm}(s_k)}{\partial W_k} = -\mathrm{sgm}'(s_k)\frac{\partial s_k}{\partial W_k} \tag{5.4}$$

We may note that

$$S_k = X_k^T W_k \tag{5.5}$$

Therefore,

$$\frac{\partial s_k}{\partial W_k} = X_k \tag{5.6}$$

Substituting into Eq. (5.4) gives

$$\frac{\partial \overline{\varepsilon}_k}{\partial W_k} = -\mathrm{sgm}'(s_k)X_k \tag{5.7}$$

Inserting this into Eq. (5.3) yields

$$\widehat{\nabla}_k = -2\overline{\varepsilon}_k \mathrm{sgm}'(s_k)X_k \tag{5.8}$$

Using this gradient estimate with the method of steepest descent provides a means for minimizing the mean-square-error even after the summed signal s_k goes through the nonlinear sigmoid. The algorithm is

$$W_{k+1} = W_k + \mu\left(-\widehat{\nabla}_k\right) \tag{5.9}$$

$$= W_k + 2\mu\overline{\varepsilon}_k \mathrm{sgm}'(s_k)X_k \tag{5.10}$$

Algorithm (5.10) is the backpropagation algorithm for the sigmoid ADALINE element. The backpropagation name makes more sense when the algorithm is utilized in a layered network, which will be studied in Chap. 6. Implementation of the algorithm (5.10) is illustrated in Fig. 5.2.

If the sigmoid is chosen to be the hyperbolic tangent function (5.1), then the derivative $\mathrm{sgm}'(s_k)$

$$\begin{aligned}\mathrm{sgm}'(s_k) &= \frac{\partial(\tanh(s_k))}{\partial s_k} \\ &= 1 - (\tanh(s_k))^2 = 1 - y_k^2\end{aligned} \tag{5.11}$$

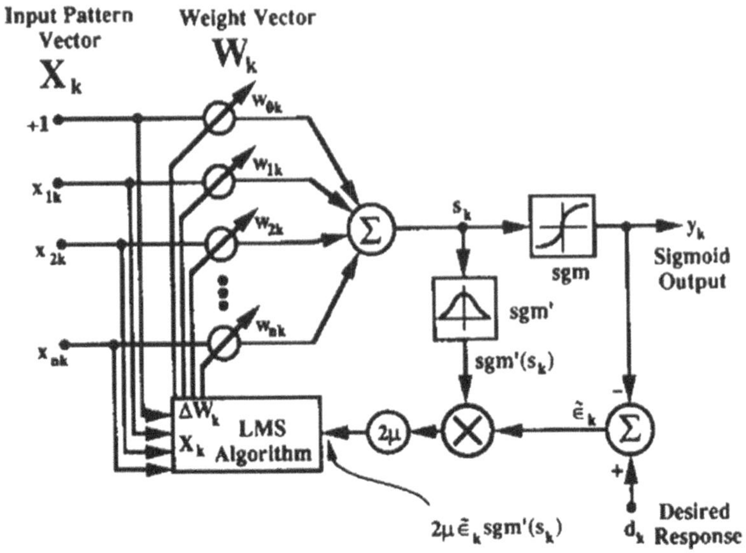

Fig. 5.2 Implementation of backpropagation for the sigmoid ADALINE element

Accordingly, Eq. (5.10) becomes

$$W_{k+1} = W_k + 2\mu\bar{\varepsilon}_k\left(1 - y_k^2\right)X_k \tag{5.12}$$

This is the LMS algorithm for the sigmoidal ADALINE.

To demonstrate the nature of the mean-square-error surfaces associated with the linear error and the sigmoidal error, a simple experiment with a two-input ADALINE was performed. The ADALINE was driven by a typical set of input patterns and the associated binary {+1, −1} desired responses. The sigmoid function used was the hyperbolic tangent function. The weights have been adapted to minimize the mean-square error of or ê. The MSE surfaces of $\mathbf{E}[(\varepsilon_k)^2]$ and $\mathbf{E}[(\hat{e}_k)^2]$ plotted as functions of two weight values, are shown in Figs. 5.3 and 5.4 respectively.

Although the above experiment is not all encompassing, we can infer from it that minimizing the mean square of the linear error is easy but minimizing the mean square of the sigmoid error can be more difficult. Only the linear error is guaranteed to have an MSE surface with a unique global minimum.

In multilayer neural networks, gradient methods generally work better with sigmoid rather than signum nonlinearities. Smooth nonlinearities are required by the backpropagation technique. Moreover, sigmoid networks are capable of forming internal representations that are more complex than simple binary codes and, thus, these networks can often form decision regions that are more sophisticated than those associated with similar signum neural networks. In fact, if a noiseless, infinite-precision sigmoid ADALINE could be constructed, it would be able to convey an

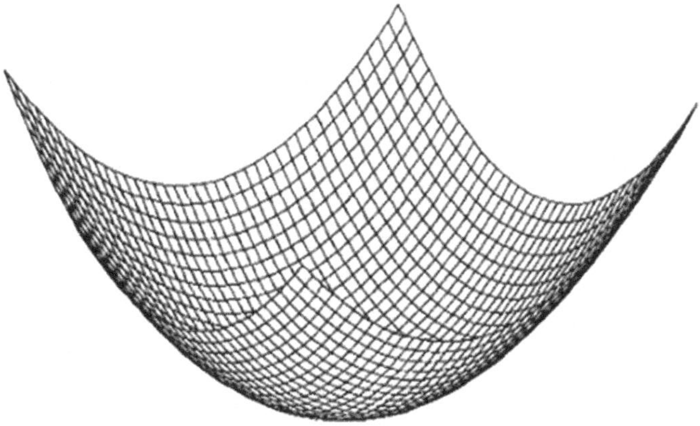

Fig. 5.3 Example of MSE surface of linear error

Fig. 5.4 Example of MSE surface of sigmoid error

infinite amount of information at each time step. This is in contrast to the maximum Shannon information capacity of one bit associated with each binary element.

5.3 Summary

An ADALINE with a sigmoidal activation function can be trained for least squares performance by using a variant of the LMS algorithm. The sigmoid LMS algorithm is more complicated than LMS but is readily executed with modern neural networks. The sigmoidal ADALINE is a major component of neural networks trained with the backpropagation algorithm.

Appendix

Use google (or your favorite internet search engine) to explore:

- Neuron activation function
- Biological neuron activation function
- Artificial neuron activation function
- Sigmoidal activation function
- MSE quadratic surface
- MSE of surface with sigmoidal activation function

Reference

1. Widrow, B., Lehr, M.A.: 30 years of adaptive neural networks: perceptron, madaline, and back-propagation. Proc IEEE **78**(9), 1415–1442 (1990). Also appears as a chapter in C. Lau (ed.) *Neural Networks: Theoretical Foundations and Analysis*, pp. 27–53, IEEE Press (1992); and in E. Sanchez-Sinencio and C. Lau (eds.) *Artificial Neural Networks: Paradigms, Applications, and Hardware Implementation,* eds., pp. 82–108, IEEE Press (1992)

Chapter 6
Backpropagation for Multi-layer Neural Networks

Abstract When training a multi-layer neural network, the desired responses of the output layer neurons are given for each input training pattern. The desired responses for the neurons in the other layers are not known. The backpropagation algorithm generates desired responses for all the neurons in the "hidden layers," the layers that feed signals to the output layer. Known errors at the output layer are backpropagated throughout the network to provide error signals for all the neurons of the network, thus enabling adaptation and learning for the entire network.

We now study rules for steepest-descent minimization of the MSE associated with an entire network of sigmoid ADALINE elements. Like their single-element counterparts, the most practical and efficient steepest-descent rules for multi-element networks typically work with one pattern presentation at a time. We will describe steepest-descent rules for multi-element sigmoid networks and backpropagation.

6.1 The Principle of Minimal Disturbance

The iterative algorithms described here are all designed in accordance with the *Principles of Minimal Disturbance: Adapt to reduce the output error for the current training pattern, with minimal disturbance to responses already learned.* Unless this principle is practiced, it is difficult to simultaneously store the required pattern responses. The minimal disturbance principle is intuitive. It was the motivating idea that led to the discovery of the LMS algorithm. In fact, the LMS algorithm had existed for several months as an error reduction rule before it was discovered that the algorithm uses an instantaneous gradient to follow the path of steepest descent and minimizes the mean squared error of the training set. It was then given the name LMS (least mean square) algorithm.

B. Widrow and E. P. Katz, *Cognitive Memory*, Springer Series on Bio- and Neurosystems 17, https://doi.org/10.1007/978-3-031-80939-2_6

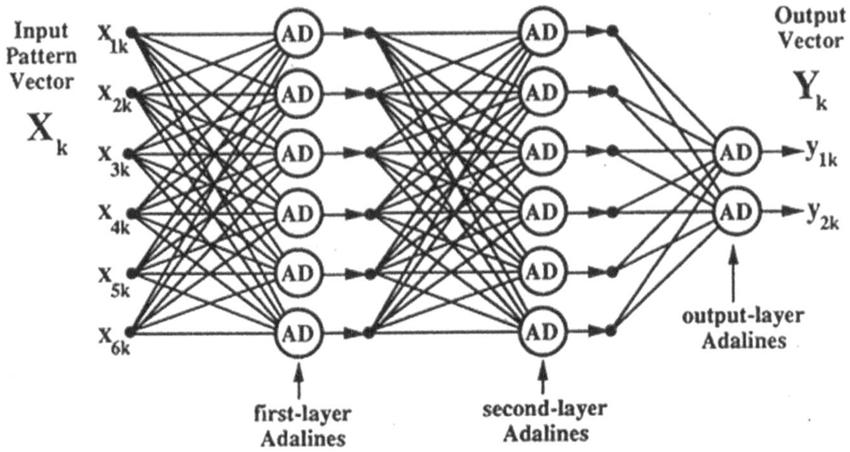

Fig. 6.1 A three-layer adaptive neural network

6.2 A Multi-layer Neural Network

The networks of the 1960's had an adaptive first layer and a fixed threshold function in the second (output) layer [1]. The feedforward neural networks of today often have many layers, all of which are usually adaptive. The backpropagation networks invented by Paul Werbos [2] and popularized by Rumelhart et al. [3] are perhaps the best-known examples of multilayer networks.

A three-layer, feedforward, adaptive network is illustrated in Fig. 6.1. It is "fully connected" in the sense that each ADALINE receives inputs from every neuron output in the preceding layer.

During training, the responses of the output elements in the network are compared with a corresponding set of desired responses. Error signals associated with the elements of the output layer are thus readily computed, so adaptation of the output layer is straightforward. The fundamental difficulty associated with adapting a layered network lies in obtaining error signals for hidden layer ADALINE's, that is, for ADALINE's in layers other than the output layer. The backpropagation algorithm provides a method for establishing these error signals. Note that layered structures only exist artificially. Biological neurons are randomly connected in three-dimensional space.

6.3 A 3-Layer Neural Network

Modern neural networks could, for some applications, have dozens of layers. A 3-layer example is shown in Fig. 6.1.

The neural elements are sigmoidal ADALINE's, designated by **AD**. The **k**th input pattern is χ_k. The outputs are represented by the vector $\mathbf{Y}_k$. The network of Fig. 6.1 is fully connected. All input vector components connect to all of the neurons of the first layer. All outputs of the first layer connect as inputs to the second layer, and so forth.

During training, input patterns χ_k's are presented and with each presentation the output vector is observed. The components of the output vector are compared with the respective components of a desired response vector $\mathbf{D}_k$ which is given in association with each input vector χ_k. The learning is supervised.

Inputs and desired responses are available to the output layer neurons. Training the output layer neurons is straightforward. Simply train with the sigmoidal LMS algorithm (5.10). If the activation functions are arctan, then Eq. (5.10) becomes (5.12).

As the desired responses at the output layer are given, the output layer neurons can be trained. But of the other layers, called "hidden layers," the desired responses of the neurons are not given. To be trained, the neurons of the hidden layers must have desired responses and their associated errors. The objective is to minimize the average of the output errors squared, the MSE of the output layer, averaged over the set of input training patterns. The backpropagation algorithm enables this. With it, errors at the output layer are back propagated throughout the network to compute errors for the hidden-layer neurons.

6.4 The Multilayer Backpropagation Algorithm

The publication of the backpropagation technique of Werbos by Rumelhart et al. [3] has unquestionably been the most influential development in the field of neural networks during the past several decades. In retrospect, this technique seems simple. Nonetheless, largely because early neural networks research dealt almost exclusively with hard-limiting nonlinearities, this backpropagation idea never occurred to neural network researchers throughout the 1960's.

The basic concepts of backpropagation are easily grasped. Unfortunately, these simple ideas are often obscured by relatively intricate notation. This notational complexity makes formal derivations of the backpropagation rule tedious. We present an informal derivation of the algorithm and illustrate how it works for the simple network shown in Fig. 6.2.

The backpropagation technique is a nontrivial generalization of the single sigmoid ADALINE. When applied to multi-element networks, the backpropagation technique adjusts the weights in the direction opposite of the instantaneous error gradient.

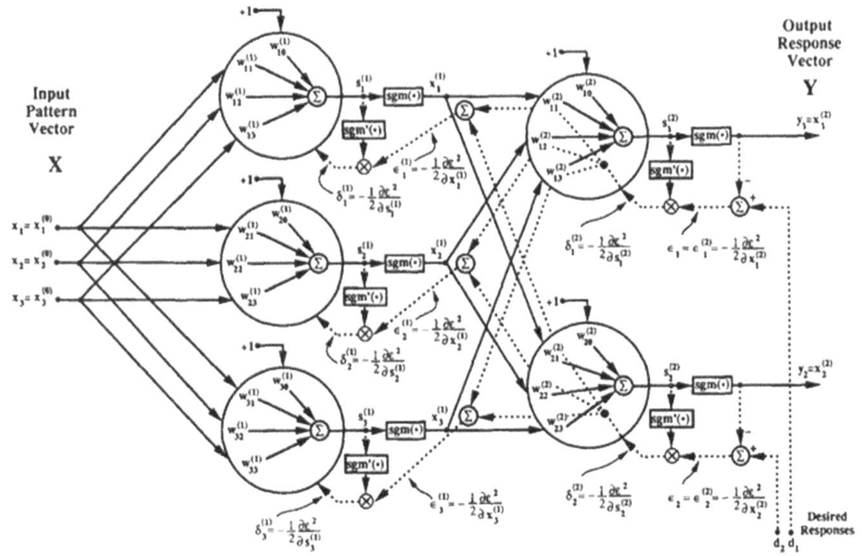

Fig. 6.2 Example of two-layer back propagation network architecture

$$\widehat{\nabla} = \frac{\partial \varepsilon_k^2}{\partial W_k} = \left.\begin{matrix} \frac{\partial \varepsilon_k^2}{\partial W_{1k}} \\ \vdots \\ \frac{\varepsilon_k^2}{\partial W_{mk}} \end{matrix}\right\} \tag{6.1}$$

Now, however, W_k is a long *m-component* vector of all weights in the entire network. The instantaneous sum squared error ε_k^2 is the sum of the squares of the errors at each of the N_y outputs of the network. Thus

$$\varepsilon_k^2 = \sum_{i=1}^{N_y} \varepsilon_{ik}^2. \tag{6.2}$$

In the network example shown in Fig. 6.2, the sum square error is given by

$$\varepsilon^2 = (d_1 - y_1)^2 + (d_2 - y_2)^2 \tag{6.3}$$

where we now suppress the time index κ for convenience.

In its simplest form, backpropagation training begins by presenting an input pattern vector X to the network, sweeping forward through the system to generate an output response vector ***Y***, and computing the errors at each output. The next step involves sweeping the effects of the errors backward through the network to associate a "square error derivative," δ with each ADALINE, computing a gradient from each δ, and finally updating the weights of each ADALINE based upon the corresponding

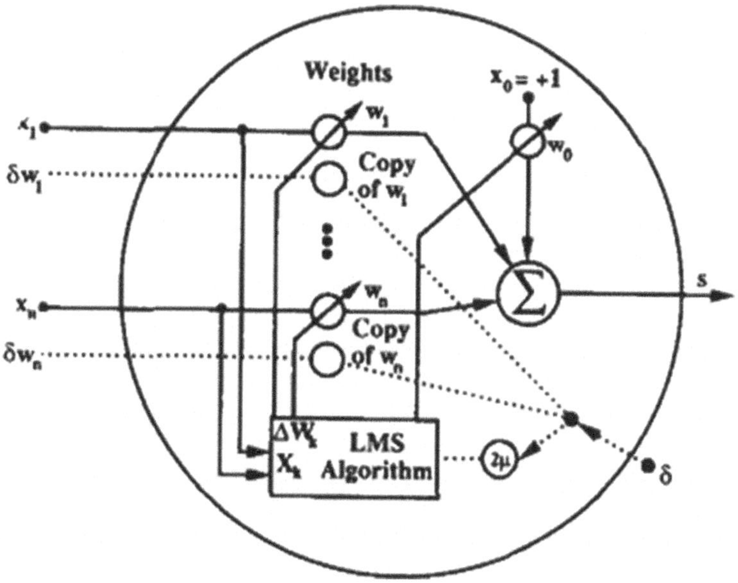

Fig. 6.3 Detail of linear combiner and associated circuitry in backpropagation network

gradient. A new input pattern is then presented and the process is repeated. The initial weight values are normally set to small random numbers. The algorithm will not work properly with multilayer networks if the initial weights are either zero or poorly chosen nonzero values.

We can get some idea about what is involved in the calculations associated with the backpropagation algorithm by examining the network of Fig. 6.2. Each of the five large circles represents a linear combiner, as well as, some associated signal paths for error backpropagation, and the corresponding adaptive machinery for updating the weights. This detail is shown in Fig. 6.3. The solid lines in these diagrams represent forward signal paths through the network, and the dotted lines represent the separate backwards paths that are used in association with calculations of the square error derivatives δ. From Fig. 6.2, we see that the calculations associated with the backward sweep are of a complexity roughly equal to that represented by the forward pass through the network. The backward sweep requires the same number of function calculations as the forward sweep.

As stated earlier, after a pattern has been presented to the network, and the response error of each output has been calculated, the next step of the backpropagation algorithm involves finding the instantaneous square-error derivative δ associated with each summing junction in the network. The square error derivative associated with the jth ADALINE in layer l is defined as

$$\delta_j^{(l)} \triangleq -\frac{1}{2}\frac{\partial \varepsilon^2}{\partial s_j^{(l)}}. \tag{6.4}$$

Each of these derivatives in essence tells us how sensitive the sum square output error of the network is to changes in the output of the associated ADALINE element.

The instantaneous square-error derivatives are first computed for each element in the output layer. The calculation is simple. As an example, below we derive the required expression of $\delta_1^{(2)}$, the derivative associated with the top ADALINE element in the output layer of Fig. 6.2. We begin with the definition of $\delta_1^{(2)}$ from Eq. (6.4).

The 2-layer network of Fig. 6.2 has three inputs and two outputs. The layers are fully connected. Referring back to Fig. 5.2, it can be seen that the error fed to the LMS algorithm of the sigmoidal ADALINE is the output error ε_k multiplied by $2\mu\text{sgm}'(s_k)$. This is the algorithm used for adaptation of the weights of the entire network. The errors of the two output layer neurons are computed this way.

The errors of the three first-layer neurons affect the errors of both neurons of the output layer. Adapting the weights of the first neuron of the first layer reduces that neuron's error. This has an effect on the error of the first neuron of the output layer which is proportional to the value of the weight coupling between the two neurons. This simultaneously has an effect on the error of the second output neuron that is proportional to the coupling weight connecting these two neurons. Thus, the error of the first neuron of the first layer is computed by summing the errors of the output layer neurons, weighted by the respective coupling weights.

The errors of the second and third neurons of the first layer are computed in a similar manner. Figure 6.2 illustrates the feedforward signal paths (the solid lines) and the backpropagation paths of the error signals (the dotted lines).

Adapting each of the ADALINES of the network with the sigmoid LMS algorithm, the mean of the sum of the output neurons' squared errors is minimized when averaged over all the training patterns. This fulfills the objective of training the network to minimize the MSE, or so it seems.

6.5 Local Optima with Backpropagation

With a multi-layer network such as, for example, the network of Fig. 6.2, the MSE at the output layer is a function of every weight value. This function is generally not a quadratic function. An example, for a simple sigmoidal ADALINE, is the function illustrated in Fig. 5.4. If with a single neuron such a surface is possible, it is clear that the MSE as a function of the weights of a network of neurons could have a great complexity. Local optima are to be expected.

In this section, we examine MSE surfaces from a typical backpropagation training problem with a sigmoidal neural network.

In a network with more than two weights, the MSE surface is high-dimensional and difficult to visualize. It is possible, however, to look at slices of this surface by plotting the MSE surface created by varying two of the weights while holding all others constant. The surfaces are plotted in Figs. 6.4 and 6.5 show two such slices of the MSE surface from a typical learning problem involving, respectively, an untrained sigmoidal network and a trained one. The first surface resulted from varying two first-layer weights of an untrained network. The second surface resulted from varying the same two weights after the network was fully trained. The two surfaces are similar, but the second one has a deeper minimum which was carved out by the backpropagation learning process. Figures 6.6 and 6.7 resulted from varying a different set of two weights in the same network. Figure 6.6 is the result from varying a first-layer weight and third-layer weight in the untrained network. Whereas Fig. 6.7 is the surface that resulted from varying the same two weights after the network was trained.

By studying many plots, it becomes clear that backpropagation will generally be subject to convergence on local optima. In spite of local optima, the backpropagation

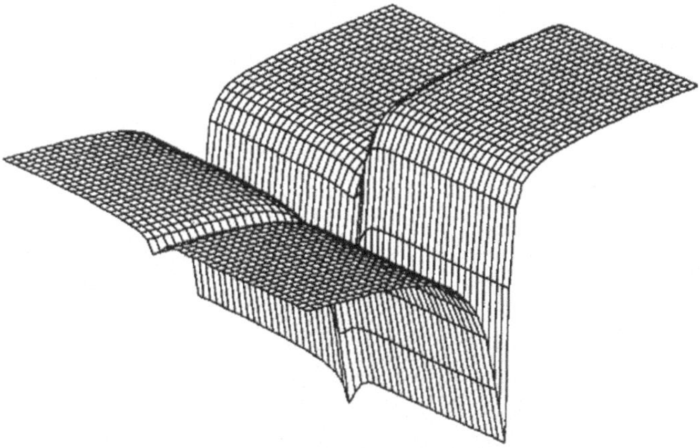

Fig. 6.4 Example MSE surface of untrained sigmoidal network as a function of two first-layer weights

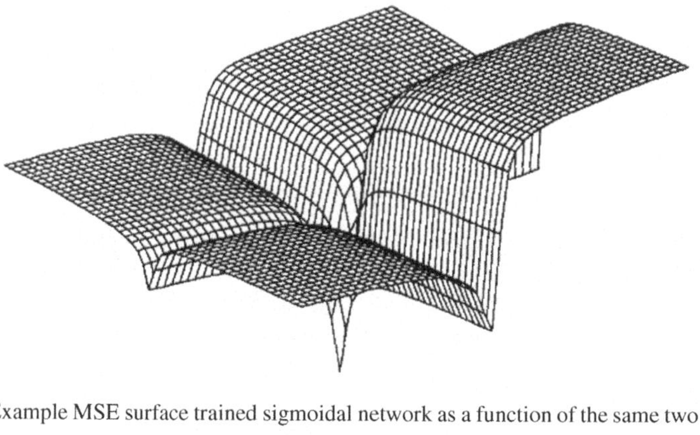

Fig. 6.5 Example MSE surface trained sigmoidal network as a function of the same two first-layer weights

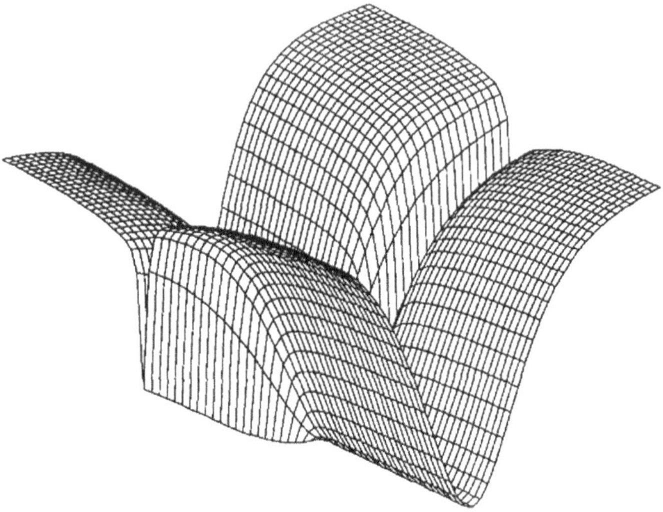

Fig. 6.6 Example MSE surface of untrained sigmoidal network as a function of a first-layer weight and a third-layer weight

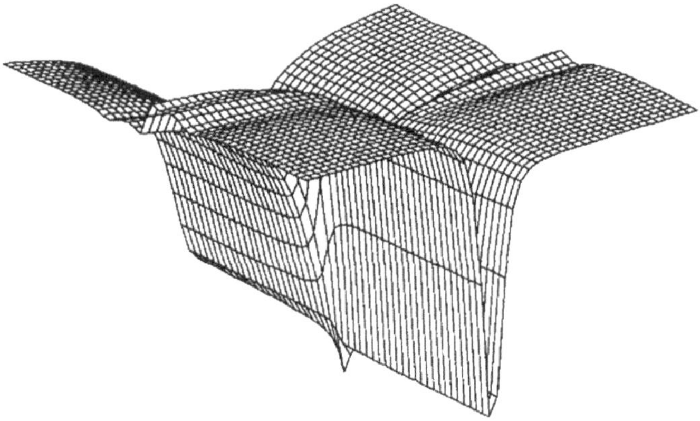

Fig. 6.7 Example MSE surface of trained sigmoidal network as a function of a first-layer weight and a third-layer weight

algorithm generally develops good solutions that are useful in practice. All neural networks of today use the backpropagation algorithm.

Solutions found by people in everyday life are usually not optimal, but many of them are useful. If a local optimum yields satisfactory performance, often there is simply no need to search for a better solution.

6.6 Summary

Backpropagation is the accepted training method for networks of sigmoidal neurons. This algorithm requires mathematical differentiation. Sigmoidal activation functions are differentiable, so networks of sigmoidal neurons can be trained with backpropagation. The object is minimization of the sum of the squares of output errors, averaged over the set of input training patterns. When the training converges, the solution obtained is generally a local optimum, not necessarily a global optimum. Nevertheless, backpropagation yields a useful solution. The world of neural networks runs on backpropagation, giving good local optima.

Appendix

Use google (or your favorite internet search engine) to explore:

- Paul Werbos
- David Rumelhart
- Rumelhart and McClelland
- Machine learning
- Multi-layer neural networks
- Backpropagation algorithm
- Backpropagation applications

References

1. Widrow, B., Lehr, M.A.: 30 years of adaptive neural networks: perceptron, madaline, and backpropagation. Proc. IEEE **78**(9):1415–1442 (1990). Also appears as a chapter in Lau, C. (ed.) Neural networks: theoretical foundations and analysis, pp. 27–53, IEEE Press (1992); and in Sanchez-Sinencio, E., Lau, C. (eds.): Artificial Neural Networks: Paradigms, Applications, and Hardware Implementation, pp. 82–108. IEEE Press (1992)
2. Werbos, P.: The Roots of Backpropagation: From Ordered Derivatives to Neural Networks and Political Forecasting. Wiley Interscience (1994)
3. Rumelhart, D.E., Hinton, G.E., Williams, R.J.: Learning representations by back-propagating errors. Nature **323**(6088), 533–536 (1986). https://doi.org/10.1038/323533a0

Chapter 7
Autoassociative Neural Networks

Abstract Input training patterns are presented to an autoassociative neural network, without desired responses. Learning is unsupervised. The training algorithm is the usual backpropagation which is a supervised learning algorithm. In an unusual manner, output desired responses are taken as the input patterns themselves. An autoassociative network is trained to produce output patterns which are identical to the input patterns that were in the original training patterns set. When presenting input patterns outside of the training set, the network outputs do not match the inputs. The difference between input and output is an error pattern. An input pattern in the training set will correspond with small errors while patterns not in the training set are associated with large errors. Once the network is trained, small errors indicate *déjà vu*, patterns that have been seen before.

Autoassociative neural networks play a major role in cognitive memory systems. Like all neural networks, autoassociative networks are presented with input training patterns in the form of training vectors. Other types of neural networks are given desired response output vectors in order to facilitate supervised learning. But with autoassociative neural networks, when input patterns are presented, no corresponding desired responses are available. The input patterns themselves are taken as the desired responses. Supervised learning takes place, but the autoassociative network is self-supervising. In a sense, autoassociative learning is unsupervised since no external desired responses are provided. But a usual supervised learning algorithm is actually used with autoassociative learning.

The autoassociative neural network is trained to produce a set of output patterns that match the set of input patterns. After training, whenever one of the training patterns is presented, the difference between the input and output patterns will be small. When a pattern which is not in the set of training patterns is presented, the network output will not match the input. The difference will be a large error. The training patterns and patterns similar to them will have small errors and thus will be detected as belonging together in an associated group, and patterns not in the training set will be "outsiders."

B. Widrow and E. P. Katz, *Cognitive Memory*, Springer Series on Bio- and Neurosystems 17, https://doi.org/10.1007/978-3-031-80939-2_7

The training set consists of patterns that have been seen before. Patterns outside the training set will be considered as not having been seen previously. When an unlabeled pattern is presented to the trained network, the question is: "Does it belong to the association or not? *Déjà vu*?".

7.1 The Architecture of the Autoassociative Neural Network

A 3-layer autoassociative neural network is diagrammed in Fig. 7.1. It has six inputs and six outputs. The input vector provides inputs to the first-layer neurons. The outputs of the first layer provide inputs to the "choke layer" in this case consisting of two neurons. The output of the choke layer provides inputs to the six neurons of the output layer.

The simplest autoassociative network has three layers. The number of outputs must match the number of inputs. The choke layer should have fewer neurons than the first layer. The number of outputs equals the number of neurons in the output layer. The number of neurons in the first layer is often made equal to the number of inputs, but this may not always be the case. Most important is that the number of outputs matches the number of inputs and that the choke layer has a smaller number of neurons.

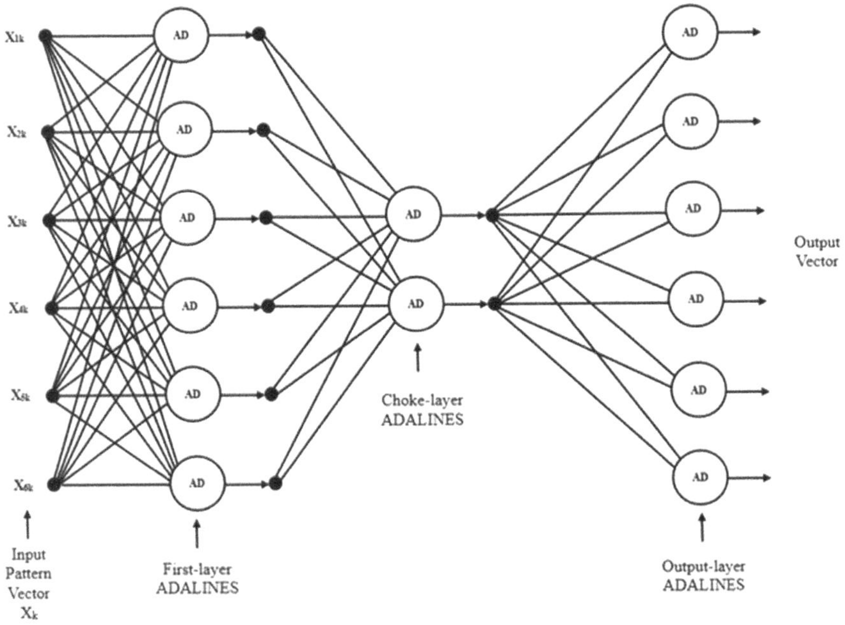

Fig. 7.1 An autoassociative neural network

7.2 More Elaborate Autoassociative Networks

The first layer could be replaced by a multiple-layered network. The output layer could also be replaced by a multilayered network. The choke layer too could be replaced by a multilayer choke network. The multiple layers increase the storage capacity of the network. The choke capacity must be less than the input capacity. Above all else, the number of outputs must match the number of inputs so that input and output vectors can be compared.

7.3 Training the Autoassociative Neural Network

Training an autoassociative network can be readily explained with the aid of Fig. 7.2. Figure 7.2a shows the conventional supervised training of a neural network. Input patterns are presented with their associated desired responses. The backpropagation algorithm could be used to train a conventional layered network.

Figure 7.2b illustrates the training of an autoassociative network. This network is represented symbolically by the "hourglass" figure. In its middle is the choke layer that has a smaller number of neurons than the other layers. The autoassociative network has input patterns, and it delivers output patterns. No external desired responses associated with the input patterns are provided. To be trained, the network must have desired responses in order to adapt and learn. The autoassociative network obtains its desired responses from its own input patterns. The difference between an input pattern and its output pattern is an error vector that can be used by the

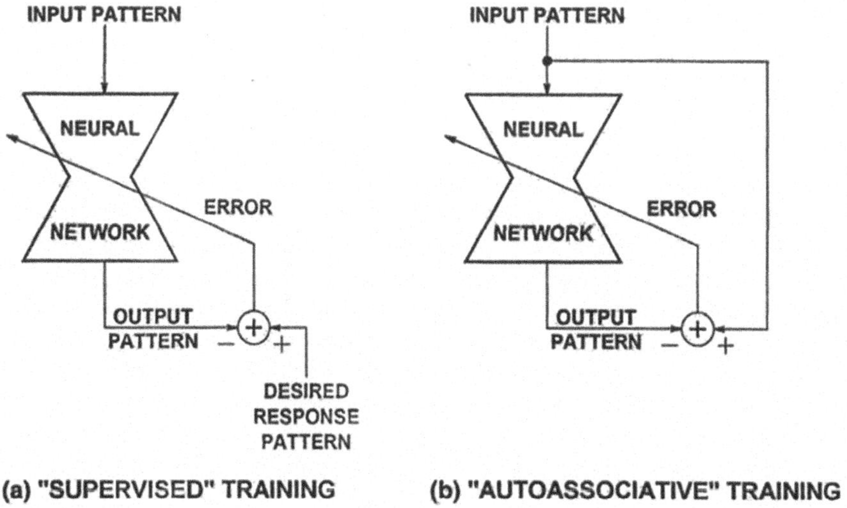

Fig. 7.2 Details of neural network training

backpropagation algorithm to affect learning. Comparing supervised learning with autoassociative learning gives one a perspective on autoassociative learning.

7.4 The Purpose of the Choke

If there is no choke layer, the number of neurons of the second layer of the network in Fig. 7.1 would be made equal to the number of neurons of the other layers. Autoassociative training would converge on a solution that would make the output patterns match the input patterns for all input patterns. (The 3-layer network would act like a straight feed through.) Many solutions exist that can do this. In this case, the autoassociative network would be useless. For all input patterns, the output pattern would be the same. But with a choke layer in place, this is not the case.

The choke layer limits the capacity of the network. The network could be trained for input–output match only over a limited number of training patterns. Once trained, the network would associate the training patterns into a cluster of patterns that experience low or zero input vs. output errors. Patterns outside this cluster would yield large errors, thus identifying themselves as not belonging to the training set. The training patterns are classified as having been "seen before". Other patterns outside the cluster are classified as having not been seen before. The magnitude of the error is the determinant. The question always is: *Déjà vu?* The autoassociative network can provide an answer.

The design of an auto associative network is still a subject of study. For a 3-layer network, the question is how many neurons to place in each of the three layers. The theoretical development is not completely finished. Experiments and simulations are the methods of design. The design is not critical, but it is important.

7.5 Déjà vu

Input training patterns are presented to an autoassociative neural network, without desired responses. Learning is unsupervised. The training algorithm is usually backpropagation which is a supervised learning algorithm. In an unusual manner, desired responses are taken as the input patterns themselves. An auto associative network is trained to produce output patterns which are identical to the input patterns that were in the set of training patterns. When presenting input patterns outside the training set, the network outputs do not match the inputs. The difference between input and output is an error pattern. An input pattern in the training set will correspond with small errors while patterns not in the training set are associated with large errors. Once the network is trained, small errors indicate training patterns that were seen before, *déjà vu*.

7.6 Summary

Autoassociative neural networks are clustering algorithms. Once trained, the network will associate all input patterns and members of the training set as distinguished from all other patterns. The clustering is unsupervised. The network learns from its own inputs. Members of the training cluster have been seen before. The network does *déjà vu*. This function is most important for cognitive memory systems.

Appendix

Use google (or your favorite internet search engine) to explore:

- Backpropagation algorithm
- Autoassociative neural networks
- *Déjà vu*

Chapter 8
The Design of a Cognitive Memory

Abstract A cognitive memory designed to record sensory input patterns, visual, auditory, tactile, etc., stores them into memory folders of conventional design. Sensory inputs are recorded only if they are "interesting." Recordings are placed into folders that have available space. Each folder is connected to a multiplexer that in turn feeds training patterns into an individual, autoassociative neural network. Training is ongoing at all times. Recorded patterns are retrieved in response to prompt inputs. These are sensory input patterns derived from the same sensors that provided signals which were recorded into the folder originally. The autoassociative neural network connects the prompt patterns to patterns recorded in the many folders of the memory system, if recorded patterns can be found that are like the prompt patterns. When there is a "hit," the contents of the respective folder are discharged onto the memory output line.

Like any memory, the cognitive memory must have both a means of data or pattern storage and a means of retrieval. We will first take up the issue of storage.

8.1 Storage

Pattern storage and a lot more is illustrated in Fig. 8.1. Sensory signals are the inputs to the memory system. The human system has a broad range of sensory inputs comprising a visual, auditory, vestibular, olfactory, tactile, and many more inputs. These inputs will be recorded in the memory. A machine memory might have analogous sensors for radar, sonar, visual, auditory, seismic, etc. inputs that could be stored into memory.

Visual inputs in the human system start with retinal signals. They are initially pre-processed by the visual cortex. The pre-processing is usually some form of compression, and the compressed signals are then available for storage. Auditory signals originate in the cochleas. From there, the signals are pre-processed by the auditory cortex and thereby undergo compression. The compressed signals are then

B. Widrow and E. P. Katz, *Cognitive Memory*, Springer Series on Bio- and Neurosystems 17, https://doi.org/10.1007/978-3-031-80939-2_8

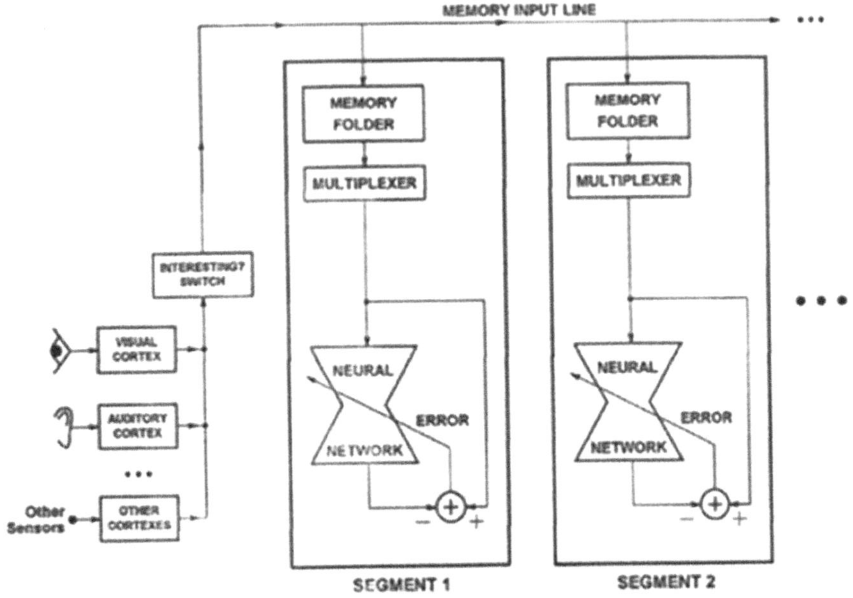

Fig. 8.1 Data recording and neural network training

available for recording into the memory system. Other sensory inputs connect to their respective cortices, become compressed, and become available for recording. Figure 8.1 illustrates this.

The processed inputs are organized as vectors. These vectors are concatenated into a very large vector. Each sensory vector maintains its separate place in the concatenated vector.

On the way to the memory system, the concatenated vector encounters a switch that is closed if the sensory signals are "interesting" and is open, blocking transmission, if the sensory signals are "not interesting." Only interesting signals can pass through the switch. For machine cognitive memory, other criteria can be developed to control the switch.

This switch in living memory is activated by mental processing. It is volitional, but more than that, it is essentially automatic. The person decides either consciously or subconsciously if the sensory inputs are interesting. If the sensory signals are interesting enough, they will pass through the "interesting" switch.

What is interesting? Perhaps a bright light, a strong color, a loud sound, a smell, or, most important, a connection is found between the current sensory inputs and previously stored inputs, stored in the cognitive memory. The concatenated signals that pass through the switch are then fed onto a memory input line that propagates throughout the memory.

The memory consists of a very large number of segments. Each segment acts independently. The memory input line carries sensory signals to all the segments simultaneously. Each segment has its own autoassociative neural network that can be trained and sensed independently, simultaneously.

The segments are analogous to the cortical columns of the living brain. The segmented structure works in parallel and can be made into a very large memory system by adding segments. Recording and access times are not affected by the memory system size, because of parallel operation.

Concatenated signals arriving at the segments of machine memory will be recorded in the first segment that has recording space. Each segment has a "red flag" that indicates if it has more recording space or not. In each segment, signals could be recorded with usual computer memory, *i.e.*, conventional RAM.

Concatenated inputs come down the memory input line and skip segments indicated by red flags. New input data is recorded at the first segment that has its red flag down. Each memory segment size determines the amount of real-time data that can be recorded there. When recording data into a given segment, recording continues until its memory is full. Then recording commences into the next segment. A given real-time episode could be recorded in many segments successively.

The memory in each segment is called a memory folder in Fig. 8.1. The memory folder of each segment is actually many memory folders, each designated for an individual sensory vector. For example, visual signals will be kept separate from auditory signals, but in a common memory folder.

Each folder is equipped with its own multiplexer and its own autoassociative neural network that enables scanning through the folder's patterns. The multiplexer scans through the folder and when finished, begins over again. The multiplexer outputs the stored patterns one at a time, repeating over and over again. Each of the sensory signals has its own multiplexer.

The patterns from the multiplexers become input training patterns for the connected autoassociative neural networks. Individual neural networks for each of the sensory signals keeps them separated. The neural networks are thus trained on the patterns stored in the segment's total memory folder. Each autoassociative network becomes an "expert" on the patterns stored in the segment's memory folder. The neural networks are trained with backpropagation, repeating over and over again, forming pattern clusters. The training never stops.

8.2 Retrieval

Every day in the life of a human, data from the surrounding world is sensed and if interesting, recorded. Day after day, the recorded database increases. A very large memory is available to the human, making all this recording possible.

The human goes about daily living and sees, hears, feels, etc. many patterns and pieces of data. These patterns and data are prompts to the memory. The human might wonder upon seeing something, "Is this interesting?" If it is, the next question is, *déjà vu*? (Have I seen anything like this before?).

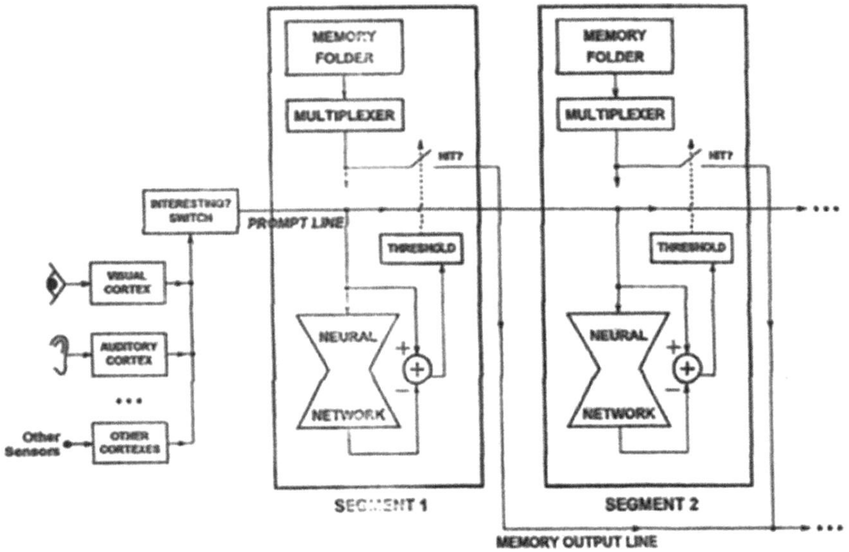

Fig. 8.2 Prompting for data retrieval

How this question might be answered can be explained by using the diagram of Fig. 8.2. This diagram shows that the sensors which observed the current input pattern or patterns in question are the same sensors that were used to create the recording in the database.

Not only are the sensors the same, but the associated cortexes are the same. The interesting switch is also the same.

The question is, regarding the current prompting input pattern, "Have I seen anything like this before?" Is there already anything in the database like it? Are there patterns in one or more folders of one or more segments that contain patterns like or similar to the current input pattern? *Déjà vu*?

Referring to Fig. 8.2, the interesting, concatenated input vector is passed through the interesting switch and is applied to a prompt line which carries prompt signals throughout the memory system. In every segment, simultaneously, the prompt signals are inputted to the trained autoassociative neural networks. The prompting input patterns are compared with the neural network output patterns, and if there is a match or a similarity, there is a hit and the contents of the respective folder are transmitted as the memory output via a memory output line, which goes throughout the entire memory. If any of the autoassociative neural networks has an error lower than a set threshold, there is a "hit" and the contents of the respective memory folder will be scanned by the multiplexer and delivered to the output line one pattern at a time.

Data pattern images are retrieved in response to a prompt input. The contents of the prompt make it possible to retrieve memory contents, regardless of where they are stored. This is a content addressable memory and a system capable of pattern recognition.

> Fundamentally, pattern recognition is about connecting a new unknown pattern with a related previously seen pattern, connecting an unknown pattern with a related pattern in memory.

If a prompt input pattern does not cause a hit in any of the segments, it would be possible to adjust the threshold settings throughout the memory, gradually raising the threshold until there is a hit. However, if stretched too far, the contents of the hit folder may be of no interest and would be discarded. This happens to people retrieving information that was not called for. One can only wonder how this happens.

As there may be as many as 10^{11} neurons and 10^{15} synapses, multiple hits would be common. A simple rule could be: output the contents of the most recently-recorded folder only. If the contents are actually the data being sought, the retrieval process could be stopped at this point. If more contents would be desired, the next most recently recorded folder would discharge its contents to the memory output line, and so forth. This method would avoid collisions on the memory output line.

8.3 Capacity

There are questions about how many patterns should be stored in a segment's memory folder, and how this relates to the design of the autoassociative neural network? The capacity of the folder should match the capacity of the autoassociative neural network. The question is:

> What is the capacity of the neural network, and what does this mean?

To begin this subject, refer to Fig. 3.1 showing the linear combiner. We can define the capacity of the linear combiner as the number of linearly independent pattern vectors which could be trained such that the MSE averaged over the training set is zero. When the MSE is zero, a set of weights exists such that the desired responses for all the training patterns can be perfectly realized. Finding the weights is similar to solving a set of **N** linear simultaneous equations with **N** unknowns. The maximum number of training patterns that can be perfectly mapped to their respective desired responses is equal to the number of weights. This number of weights is equal to the linear combiner's capacity.

The diagram of a sigmoidal ADALINE can be found in Fig. 5.1. Of interest is the capacity of the sigmoidal ADALINE, a linear combiner with a nonlinear output. The sigmoid function is monotonic. For every value of the sum, and the linear output s_k, there is a unique value of the sigmoidal output y_k. The number of training patterns that can be mapped to match their desired responses will be the same as for the linear combiner itself. The sigmoidal ADALINE capacity is therefore equal to the number of its weights. Note that the input pattern vectors and the sigmoidal outputs could be analog (continuous in amplitude) or digital (binary).

We now take a big leap and try to determine the capacity of an entire autoassociative neural network. Figure 7.1 shows a 3-layer autoassociative network. The neurons of the output layer all have the same number of weights where the number of inputs to these neurons is equal to their number weights. The capacity of the output layer neurons is equal to their number of inputs. Their number of inputs is equal to the number of neurons in the choke layer. Thus, the capacity of the network cannot exceed that of the number of choke layer neurons.

The choke-layer neurons have capacity equal to the number of their inputs, i.e. equal to the number of first layer neurons. The number of neurons of the first layer could be made large so that this would not be limiting the network capacity. Also, the first layer neurons could have many weights and many inputs, so there would be no limitations on network capacity from the first layer.

The output layer would have the same number of outputs as the network has inputs. The number of neurons in the output layer thereby determines the number of weights of the first layer neurons. The number of neurons in the output layer determines the capacities of the first layer neurons.

To summarize, the capacity of the first layer is determined by the number of neurons in the output layer. The capacity of the second layer, the choke layer, is determined by the number of neurons in the first layer. The capacity of the output layer is determined by the number of neurons in the choke layer. If the number of neurons of the first layer and the output layer are chosen to be large, and the number of neurons in the choke layer is chosen to be small, then the capacity for the entire network will be determined by the number of the neurons in the choke layer. The network capacity will be limited by the output layer capacity which is equal to the number of neurons in the choke layer.

To clarify, we ask the question,

What is the meaning of the capacity of the autoassociative network?

The network capacity is equal to the number of training patterns that could be perfectly trained in with a zero MSE. At capacity, sets of weights throughout the entire network exist that if chosen would cause the MSE to be zero. Anyone of those training patterns would be a "hit" pattern. Those patterns not in the training set would be "no-hit" patterns.

The number of patterns recorded in each memory folder is chosen to be a practical number. The number of choke neurons is chosen to match the capacity of the memory folder. The match need not be perfect. An approximate match would be fine. When a cognitive memory is prompted, the MSE in a hit segment would generally be small and not necessarily be zero.

Part 3 follows. Several applications of machine cognitive memory are presented.

8.4 Summary

A machine cognitive memory is designed having a large number of segments. The size of the memory is determined by the number of segments. Each segment has a conventional pattern store and autoassociative neural networks. The autoassociative networks are trained to do clustering with patterns or data permanently stored in folders of the segments. During retrieval, the autoassociative networks can connect an input prompt with stored similar patterns. The cognitive memory is an unsupervised data processor that can perform memory functions and pattern recognition. A machine cognitive memory can be constructed by using conventional digital computer hardware.

Appendix

Use google (or your favorite internet search engine) to explore:

- Memory storage
- Memory retrieval
- Prompt for memory retrieval
- Multiplexer
- Long-term memory
- Cortical columns

Part III
Applications of Artificial Cognitive Memory

Part III illustrates how the cognitive memory concept can solve practical engineering problems. Memory and pattern recognition intertwine to form a central idea. That is, the key to pattern recognition is memory. Artificial cognitive memory is in many respects like human memory. It has many uses, among them is pattern recognition.

There are many other ways to solve pattern recognition problems. The purpose of Part III is to demonstrate that the cognitive memory approach is viable and worthy of deeper study. Major issues are accuracy, data requirements, and computational requirements. Artificial cognitive memory is not biological. But by studying its behavior, insight might be gained regarding the behavior of nature's cognitive memory.

Chapter 9
Chinese Characters

Abstract A machine cognitive memory system can be used to automatically recognize Chinese characters. These character images may be noisy or partially obscured. But they can in many cases still be identified.

Chinese characters are pictorial, representing meaning in the form of images. There are thousands of Chinese characters.

9.1 Recognition of Chinese Characters

The machine cognitive memory system was used to recognize Chinese characters. A database of 20,000 Chinese characters was available. Of these, 10,000 randomly chosen characters from this database were used to train an auto-associative neural network. When the trained network was tested, all the training patterns were correctly identified as belonging to the training set. Of the remaining Chinese characters not used for training (non-trained patterns), 97% were correctly identified as not having been trained. Of the non-trained patterns, 3% gave false-positive indications that they belonged to the training set. This situation arises because of many close similarities between certain Chinese characters.

Further tests were performed with a group of 1000 randomly selected Chinese characters. They were trained into cognitive memory. Figure 9.1a shows one of the trained-in characters. Figure 9.1b shows the output of the auto-associative neural network. Ideally, the input pattern should be perfectly reproduced in the neural network output. In this case, the output was close to perfect, with an MSE of 0.25%. Certainly, the Chinese character of Fig. 9.1a could be recognized and identified.

B. Widrow and E. P. Katz, *Cognitive Memory*, Springer Series on Bio- and Neurosystems 17, https://doi.org/10.1007/978-3-031-80939-2_9

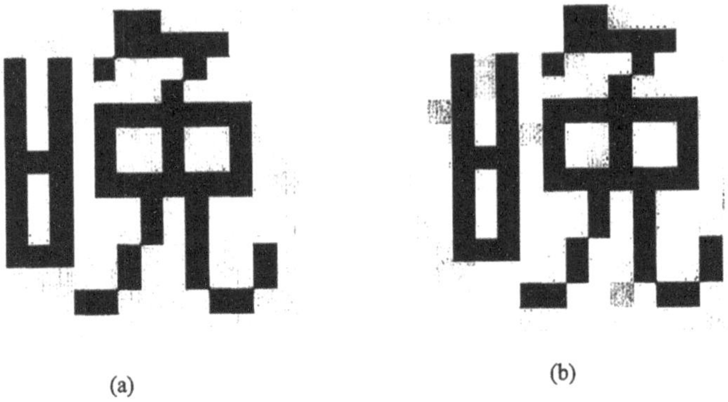

Fig. 9.1 Chinese characters

9.2 Distorted Chinese Characters

The Chinese character of Fig. 9.1a was then partially obscured, as shown in Fig. 9.2a. This obscured pattern was then inputted to the already-trained neural network, trained with perfect character images. The output, shown in Fig. 9.2b, does not perfectly match the original input of Fig. 9.1a. This is to be expected. There is an MSE of 26.82%. In spite of the large error, the pattern of Fig. 9.2b contains much of the original information. One might guess which character it is. The distorted character could probably be recognized, but one would need to accept a high MSE when calling this a hit. Accepting hits with a high MSE in some cases would cause retrieval of unintended patterns. This is common with human memory.

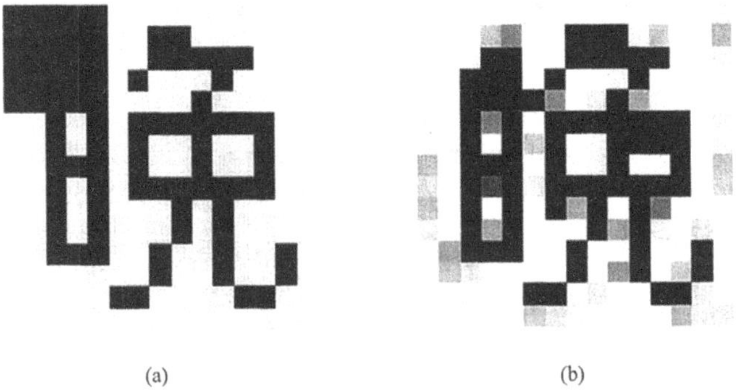

Fig. 9.2 Partially obscured Chinese characters

All of the 10,000 trained Chinese characters gave almost perfect responses at the auto-associative neural network output. The partially obscured patterns gave surprisingly good outputs from the neural network, which can be observed in the examples of Figs. 9.3 and 9.4.

The implications of this test are that it would be possible to read Chinese printing even if the printing was not perfect, given a high MSE threshold. In a cognitive memory, the Chinese character database would be organized such that a single character, complemented with random noise to capacity, would be stored in each segment. Thus, 10,000 characters would occupy 10,000 segments. When prompted with a new incoming character, the memory would find the segment for that character and would output the contents of the segment's memory folder containing the character's identification. Other ways to do this are possible.

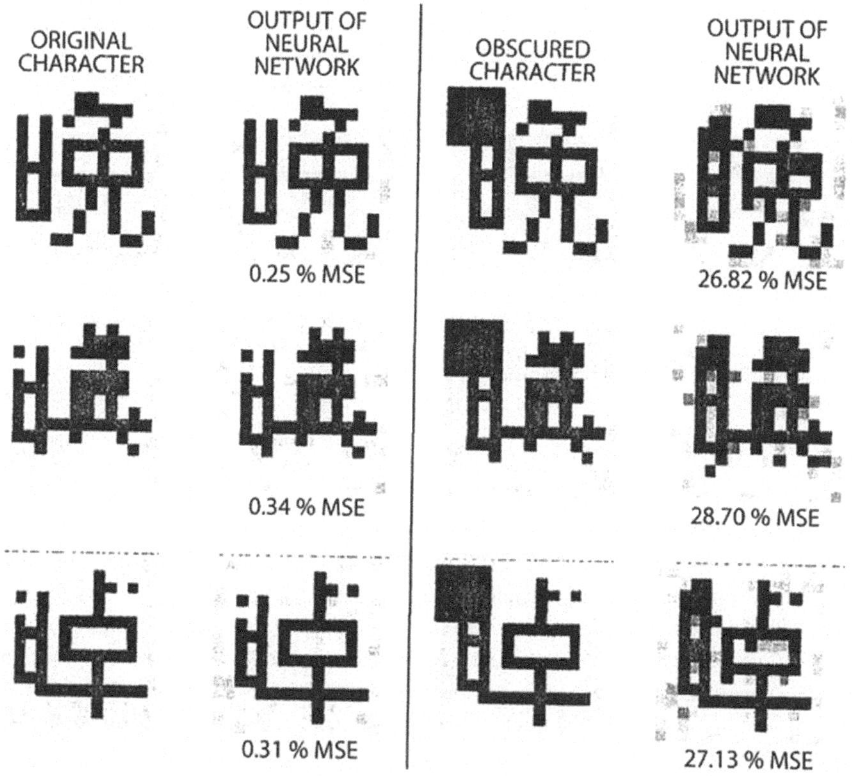

Fig. 9.3 Partially obscured Chinese characters

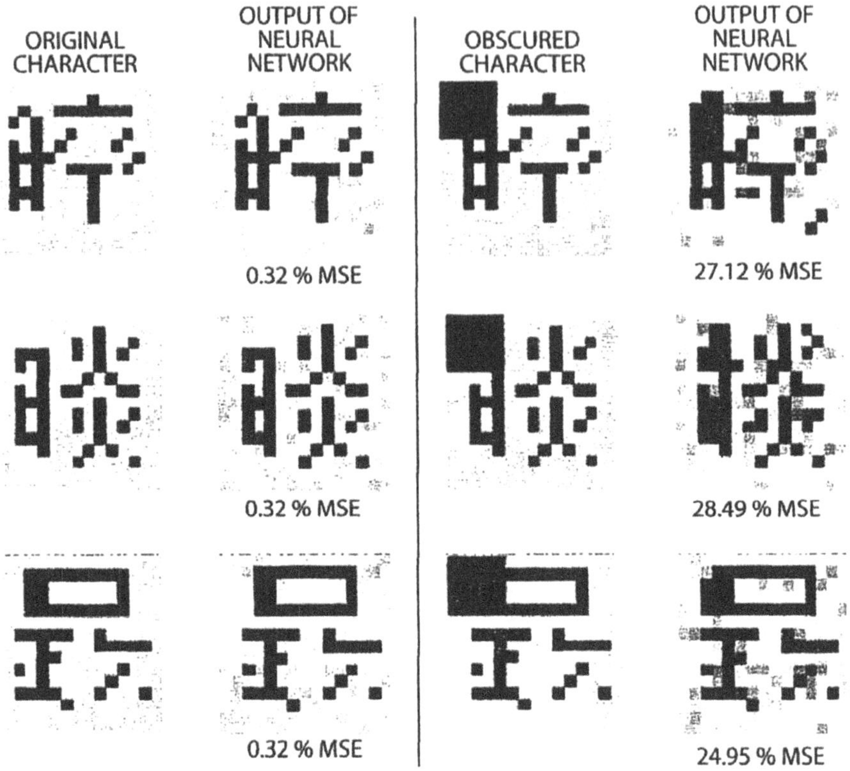

Fig. 9.4 Partially obscured Chinese characters

9.3 Summary

As described, 10,000 Chinese characters were trained into an auto-associative neural network of a single segment. Testing the trained system by prompting with any of the training patterns gave a perfect indication of *déjà vu*. Testing with Chinese characters that were not in the training set, there was false indication of *déjà vu* with approximately 3% of the characters. The auto-associative network was able to indicate whether or not an unknown prompt pattern was seen before. It was also able to identify the actual character, making the right choice out of thousands of possibilities. Recognition of noisy characters can be done. But this will require a higher MSE threshold, and would result in a higher state of misclassification. Partially obscured characters could also be recognized with a higher MSE threshold, but there would also be a higher misclassification rate.

Appendix

Use google (or your favorite internet search engine) to explore:

- Character recognition
- Noisy character recognition
- Partially obscured character recognition
- Chinese character recognition
- One out of many selection
- Noisy Chinese characters.

Chapter 10
Diego Garcia

Abstract The U.S. Air Force has a base on the small island Diego Garcia, located in the Indian Ocean. This island is about two miles long and is narrow. There is a runway on the island. A satellite photo of Diego Garcia can be seen in Fig. 10.1. The problem is to identify the aircraft parked on the tarmac on the two sides of the runway. In the satellite photos, there are three types of aircraft: B52 heavy bombers, KC135 tankers for in-flight refueling, and B2 stealth bombers. The images of the aircraft were taken through foggy conditions, making identification challenging. Sunlight striking the aircraft gave different perspectives and shadows between those on one side of the runway and those on the other side. One sample of the B52 images was taken on the other side. The same was taken for the KC135's. Only one sample image of the B2 was taken. These samples were labeled and used to train a machine cognitive memory. After training, the memory system was able to perfectly identify all the aircraft images, both training patterns and patterns not in the training set. The recognition process was tolerant of the variations in the foggy images.

Diego Garcia is a small island in the Indian Ocean. The Portuguese were the first Europeans to discover it. The French later settled it, followed by the British, and today it is a British-owned island. It is used by the British navy, the U.S. Navy, and the U.S. Air Force. It is located about 1,000 miles from the nearest continent.

10.1 Aircraft on Diego Garcia Island

The U.S. Air Force uses this island as a base primarily for B52 and B2 bombers and KC135 air refueling, tanker planes. A satellite photograph of Diego Garcia, Fig. 10.1, was obtained from a website which shows aircraft parked on a tarmac area adjacent to the island's principal runaway. The cognitive memory system was used to locate and identify the parked aircraft. Being able to do this allows constant automatic surveillance of Diego Garcia, and of any other airfield in the world.

In Fig. 10.2 is a photo of a B52 intercontinental bomber.

B. Widrow and E. P. Katz, *Cognitive Memory*, Springer Series on Bio- and Neurosystems 17, https://doi.org/10.1007/978-3-031-80939-2_10

Fig. 10.1 Diego Garcia Island

Fig. 10.2 U.S. Air Force B52 conventional bomber in flight

A photograph of the KC135 tanker is in Fig. 10.3.

Figure 10.4 is a photo of the B2 stealth bomber.

Figure 10.1 is a satellite photograph of a major portion of Diego Garcia Island. Zooming in, the aircraft parked near the main runway are clearly visible in Fig. 10.5. The smaller four engine aircraft are KC135 refueling tankers. The larger aircraft, each with four engine pods, are B52's. The B52 actually has eight engines, two engines

Fig. 10.3 U.S. Air Force KC135 tanker refueling a fighter jet in flight

Fig. 10.4 U.S. Air Force B2 stealth bomber in flight

per pod. However, that much detail cannot be seen in the photographs. Looking at the upper row of aircraft, counting from left to right, there are two KC135's, an open space, two B52's, what appears to be a large grease spot on the tarmac (it is actually a B2 stealth bomber), two more B52's, another big grease spot which is another B2, and finally two buildings that are B2 shelters. Looking at the lower row of aircraft, from left to right, there are two KC135's, a space, two more KC135's, a large space, three B52's, another large space, and finally a KC135.

A zoomed-in picture of the five KC135's is seen in Fig. 10.6. Two are from the upper row, and three are from the lower row. The upper row and lower row airplanes were parked in opposite directions, and the shadows and perspectives are correspondingly opposite for these planes. The image of an upper row KC135 is not exactly a rotated version of the image of a lower row KC135. This is because they are somewhat different images. The same effects can be seen regarding shadows and

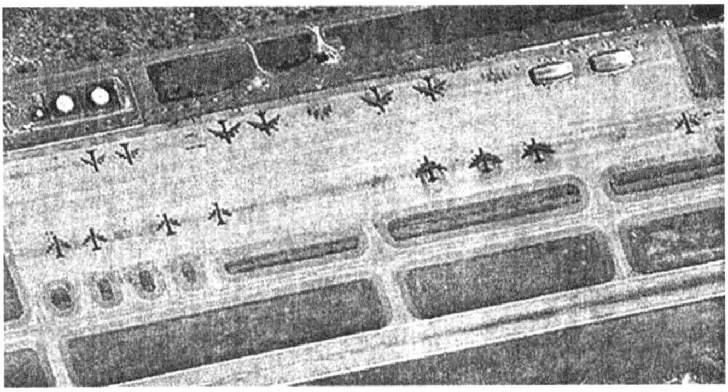

Fig. 10.5 Parked aircraft on Diego Garcia

perspectives for the B52's. In Fig. 10.7, four zoomed-in images of B52's can be seen. Visible in these images are the dark shadows from the wings and fuselage.

Fig. 10.6 Five KC135 tanker aircraft

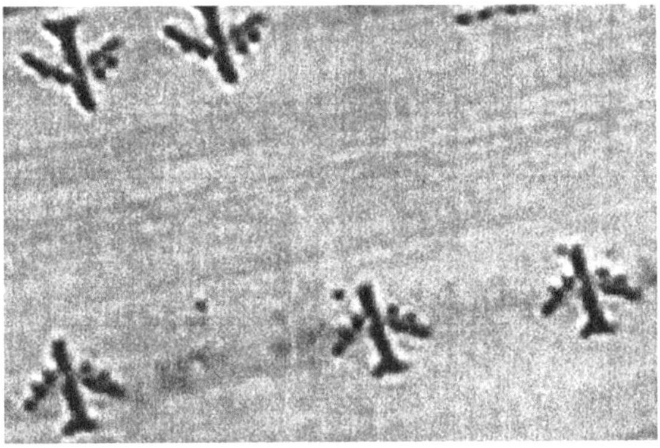

Fig. 10.7 Four zoomed in B52 bomber aircraft

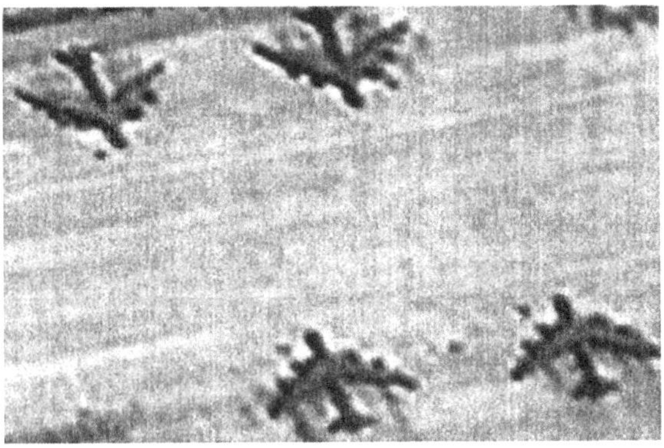

10.2 Recognition of Aircraft from Foggy Images

Two of the B52 images from Fig. 10.7 were selected, one from the upper row and one from the lower row. These were treated as two separate images because of the differences in perspective and shadow. More images were created by translating these images left–right and up-down, so that their centers were moved over a (5×5 pixel) grid. The result was $25 \times 2 = 50$ different B52 patterns. All 50 of these patterns were recorded in a single memory folder in memory segment 1. In addition, the memory folder had the identification of the aircraft, e.g., B52. All of the airplane images in the memory folder were trained into an autoassociative neural network. Separate memory folders were used to hold images of KC135 and B2 aircraft, respectively.

Images from the satellite picture of Fig. 10.4 taken within the "mind's eye" (a square area somewhat bigger than the B52 image) acted as prompts. The mind's eye square was programmed to scan (as in television horizontal line scanning) the entire satellite photograph to look for hits. In each scanned position, the mind's eye was rotated in 5-degree increments, looking for a hit, intending to minimize the mean-square-error of the autoassociative neural network. When a hit took place, the memory segment was searched for correspondence with the mind's eye image that caused the hit, the memory folder was noted, and the identification of the aircraft was downloaded from the memory folder. The X–Y physical location of the aircraft was determined from the X–Y image coordinates of the center position of the mind's eye square.

A hit corresponded to the mean-square-error (MSE) of the auto-associative neural network being very low. As the mind's eye scanned and came upon the image of a B52, it scanned this image up and down, left and right, making many hits because the B52 patterns were trained with 25 different positions. For each position of the mind's eye, many rotations with 5–degree increments were tried to find the best fit. After getting many hits and identifying them as B52, the computer screen showed "B52 Bomber." When the mind's eye scanned over a KC135, various positions and rotations gave a very low MSE. Then the computer screen indicated "KC135 Tanker." The detection error threshold was pre-set but could be relaxed or tightened,

as described above. Figure 10.8 shows the aforementioned "grease spot," which is actually a B2 bomber. The image within the "mind's eye" rectangle was stored in the cognitive memory and later recognized as a B2. After recording a hit for this B2 bomber, the memory folder storing B2 images ignored the B52s, but correctly identified the second B2 bomber. The image quality of the second B2 was much poorer than that of the first B2, which was the image used to train the system. In spite of this, the second B2 was detected, located, and identified. There was a lot of tolerance in matching the prompt pattern to the corresponding trained pattern.

The cognitive memory could have been trained to locate and identify many other types of objects on that island in addition to the aircraft, scanning over the entire aerial photograph. Once the cognitive memory has been trained to seek out and identify the types of objects of interest, scanning and detections can be made at very high speed with a parallel hardware implementation.

Fig. 10.8 Two B52 bombers and a B2 "grease spot."

10.3 Summary

Images were taken from a satellite photo of the U.S. airbase on the island Diego Garcia in the Indian Ocean. The images of the three different aircraft were identified by a cognitive memory system. The aircraft were parked on the tarmac on both sides of a runway. There were variations in perspective, lighting, and shadows for these aircraft, major differences between aircraft on one side of the runway and those on the other side. The memory system was trained with a very small number of images of each type of airplane. Then testing was performed on images not seen before. A square window was placed over the image of one B52 airplane. The window was used to scan the satellite photo, looking for any aircraft. It was necessary to rotate and translate the square window, in small steps when looking for aircraft. When a hit was found, the aircraft's identity became known.

Appendix

Use google (or your favorite internet search engine) to explore:

- Satellite images of aircraft
- Aircraft recognition on tarmac with noisy images
- Diego Garcia Island
- USA—B2 aircraft
- USA—B52 aircraft
- USA—KC135 aircraft.

Chapter 11
Face Recognition

Abstract A method for face recognition using a cognitive memory system is described in this chapter. A photograph was given showing many faces. Background clutter was present in addition to background noise. The problem was to identify all of the faces. A separate photo of an individual was also available. A 20 × 20 pixel low resolution version was created from the individual photo. From this, a large number of low resolution images were created by translation, rotation, scaling, and varying brightness and contrast. The 20 × 20 images looked like the images of a person's face, but with low resolution, it would be impossible to identify the person. A low resolution cognitive memory was trained with the low resolution images. The low resolution images served as templates. The templates scanned the multi-face photo to locate and identify human faces. High resolution 50 × 50 images were then made from the detected facial images. A larger number of images were created from the detected images by translation, rotation, scaling, etc. They were presented to a high resolution cognitive memory that had been trained with labeled single photos of the individuals. Recognition was accomplished.

11.1 Training Process

Recognition of human faces can be accomplished by making use of cognitive memory.

Referring to Fig. 11.1, three photographs of Bernard Widrow can be seen. These were used for training (all the photographs used in this application were converted from color to grayscale).

Figure 11.2 is a test photograph in which Juan Carlos Aragon,[1] Victor Eliashberg, and Bernard Widrow can be seen together. The problem was to automatically scan the test photograph, locate and identify the face of Widrow, and reject background objects and the faces of Aragon and Eliashberg. Widrow's face, taken from three separate photos, had been used for training; photos of Aragon and Eliashberg were not used for training.

[1] The experiments described in this chapter were performed by Juan Carlos Aragon.

B. Widrow and E. P. Katz, *Cognitive Memory*, Springer Series on Bio- and Neurosystems 17, https://doi.org/10.1007/978-3-031-80939-2_11

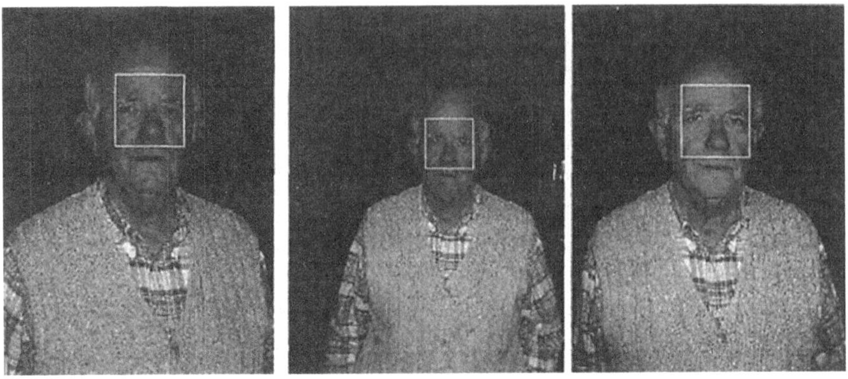

Fig. 11.1 Three training photographs of Dr. Widrow

Fig. 11.2 The test photograph

Referring to the three training photographs of Widrow in Fig. 11.1. Each photograph can be seen with a square window "mind's eye" superposed. Each window was digitized to have 50×50 pixels. These square windows were manually placed on the facial images. They were made to have various sizes, three different scalings, for each of the photographs. The placements and the sizes were not critical. Most of the windows were set to capture the two eyes and the nose. Each window was rotated clockwise and counterclockwise over seven different angles of rotation, spaced two degrees apart. The image within each window was adjusted in brightness for two different levels. Thus the number of training patterns obtained from each photograph from within its window was $3 \times 7 \times 2 = 42$ patterns.

Each pattern was digitized to have 50×50 pixels. This was made possible by interpolating the original photograph and re-sampling for each of the windows. Since there were three different training photographs, the total number of training patterns was 126. All of these training patterns were stored into a single memory folder of the first memory segment, and were labeled "Widrow."

The 126 training patterns were first converted to 50×50 vectors. The 126 vectors were then used to train the autoassociative neural network. The training objective was to make the error, the differences between the input pattern vector and the output pattern vector, very small. After training was completed, the largest mean-square-error (MSE) among all of the training patterns was of the order of 0.25% MSE.

The number of pixels for each training pattern was $50 \times 50 = 2500$ pixels. The autoassociative neural network was designed to have 1800 neurons in its first layer. Each of these neurons had 2500 weights. The network's second layer consisted of 1200 neurons, each with 1800 weights. The third layer (which was also the output layer) had 2500 neurons with 1200 weights each. The number of inputs for this autoassociative network was 2500, and the number of outputs was 2500. The total number of neurons was 5500, and the total number of weights was 9,660,000. The choices of number of layers, number of neurons, and number of weights was not critical. Wide variations in these parameters gave equivalent results.

This neural network was the high-resolution version of two networks. A low-resolution autoassociative neural network, was a separate structure and was independently trained. Low-resolution patterns having 20×20 pixels were trained from a photograph of a person's face in a manner similar to that used to obtain the high-resolution patterns. The low-resolution autoassociative network had 400 neurons in its first layer, each with 400 weights. There were 250 neurons in the second layer with 400 weights each. The third and output layer had 400 neurons with 250 weights each. The total number of neurons was 1050, and the total number of weights was 360,000. Once again, wide variations in these parameters gave equivalent results. The low and high-resolution networks were similar in structure and training methods, but were used for different purposes. Once these networks were trained, they were used to meet the pattern-recognition objectives.

11.2 The Recognition

Inspecting by eye 20×20 images of photographs, one could see human faces but could not identify the individuals. Inspecting 50×50 images one can readily identify the individuals. The low-resolution network was used to locate the various faces in the test photograph of Fig. 11.2. In Fig. 11.3, a window is scanned over the photograph like a horizontal TV scan: left to right, then to the next lower line, then left to right again, then lower again, then left to right again, etc. In each position, the pattern in the window was applied to the low-resolution network. The window was rotated in 2-degree increments, 8 different window sizes, and 16 different brightnesses. When the best combination of these variables was found (minimum mean-square-error, MSE) for each window position, the MSE was compared with a pre-set threshold. When the window was centered on one of the faces, the best combination of rotation, scaling, and brightness resulted in a very low MSE, and the low-resolution autoassociative network thereby gave a hit indication, and a face had been located (Refer to Fig. 11.3).

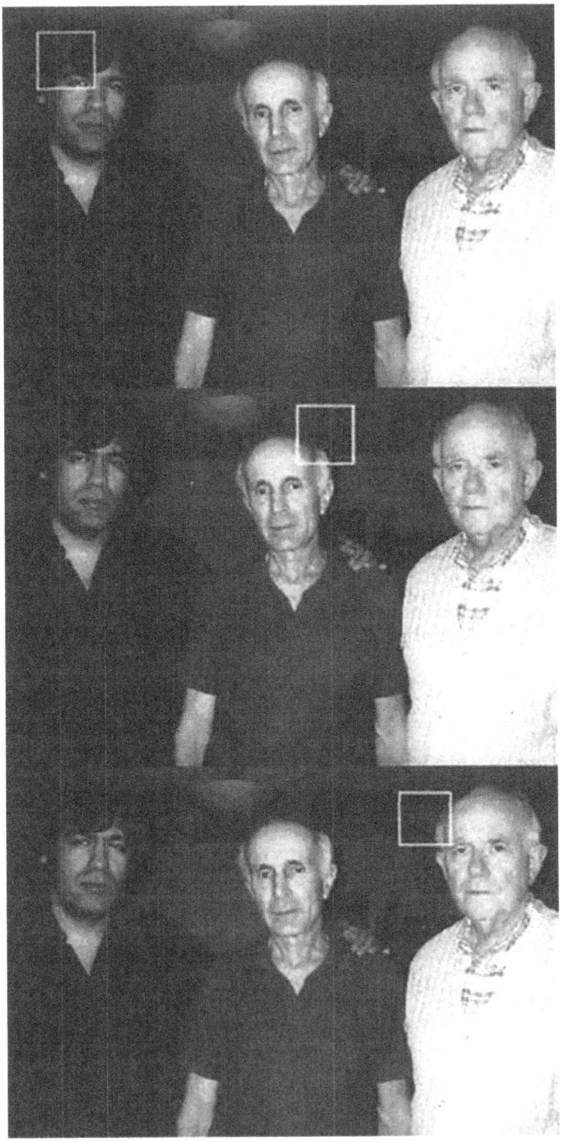

Fig. 11.3 Scanning of the test photograph

Once a face was detected (Fig. 11.4), the system switched over to the high-resolution network. The initial center of the high-resolution window was taken from the center of the low-resolution window corresponding to the low-resolution hit. The same low-resolution network was equally effective in detecting all three faces in the test photograph.

This is the advantage of the low-resolution approach, to localize the approximate location of the faces before performing high-resolution sensing everywhere. The 20 × 20 patterns allow the detection of a face, but not a recognition of whose face it was.

In Fig. 11.5, one can observe some of the search process with the high-resolution network. Windows were constructed of 10 different sizes over 15 different left–right, up-down positions, over seven different rotation angles spaced two degrees apart, with 40 different brightness levels. If the face and its variations did not cause a hit with the high-resolution neural network, the face was not recognized.

In Fig. 11.6, another face was detected by the trained low-resolution network. Scanning over the face with the trained high-resolution network, varying left–right, up-down position, rotation, scaling, brightness, and contrast as before, a hit occurred. This gave a hit pattern corresponding to one of the training patterns. All training patterns were stored in a memory folder. The memory folder was then searched to find this training pattern by comparing the hit pattern on a pixel-by-pixel basis with each of the patterns in the memory folder. The error was minimized in comparison with one of the patterns in the memory folder. This pattern was previously labeled "Dr. Widrow". Therefore, the face was identified as belonging to Widrow.

In Fig. 11.8, another face was detected, and with high-resolution examination, there was no hit. Therefore, the face was not recognized.

A relaxation method was used to obtain the best combination of parameters: translation, rotation, scale, contrast, and brightness. This process starts with the X–Y coordinates of the center of the square window that generated the hit in the low-resolution face detection process described previously. The next-step was to scan locally with the trained high-resolution network and to initiate an optimization cycle by setting initial conditions for four parameters, which were the X–Y coordinates of the square window's center (these were the coordinates from the previous hit), an initial value for the square window's size, and an initial value for the low-resolution brightness level to be used in the optimization process with the trained high-resolution network.

After the initial conditions are set, the next step is to perform a shift in x-coordinate optimization. This optimization process performs a left–right translation looking for the minimum MSE while keeping constant the initial y-coordinate and the initial brightness level. This process is very similar to the one described for the aircraft identification process. However, in this case, the attempt is made to try three scaling windows at every position of the left–right shift, looking for the optimal window (the window that generates the lowest MSE at each point).

The purpose of this was to ensure that the correct scaling was obtained for face identification at every position in the process. After the shift in x-coordinate optimization was completed, the shift in y-coordinate optimization was started. Scaling window optimization at every position of the up-down shift was performed. Finally brightness optimization was performed ending one optimization cycle.

The MSE was recorded after the four parameters mentioned above were optimized. These optimized parameters would become the new initial conditions for another optimization cycle. After several optimization cycles were completed, the configuration was located close to the global minimum of the four dimensional (due to the four parameters being optimized) on a performance surface. At this point, the image that results from applying the optimized parameters to a comparison process is identified. If the MSE is lower than a pre-set threshold, there will be a hit and the contents of the folder containing the hit photograph will be outputted and the hit photograph will be identified.

In addition to creating prompt patterns directly from input patterns or photographs, one could process the input patterns and/or the photographs or images, such as for example by rotating, translating, scaling, varying or adjusting brightness, varying or adjusting contrast, varying or adjusting color characteristics, varying or removing distortion, distending, compressing, squeezing, shearing, warping, or other processes that alter the tonality or geometry of the prompt patterns or photographs, when attempting to find a "hit" with the autoassociative neural network.

11.3 Percent MSE Curves for Face Identification

With the rest of the parameters optimized (scaling, brightness and y position), Fig. 11.10 shows how the MSE of the trained high-resolution autoassociative neural network increases as left–right translation was changed from optimal.

Figure 11.11 shows how the MSE increases as up-down translation is changed from optimal. Figure 11.12 shows percent MSE as the window size is zoomed in and out relative to the optimal window size. Figure 11.13 shows how rotation affects MSE, and Fig. 11.14 shows how pixel intensity, or brightness, affects the MSE. These curves turn out to be quite smooth, convex, and easy to search with relaxation methods. That is one of the great advantages in working with an autoassociative, layered neural network. Another advantage in working with an autoassociative neural network is that multiple photographs of an individual can be used to generate training patterns, and thus, improve recognition accuracy over training with a single photograph.

11.4 Recognition with a Group of Individuals

In another set of experiments, two photographs of each of 144 persons were available. A single photograph of 44 of these persons was selected for training. Each training photograph was varied by rotation, translation, scaling, brightness and contrast, as with the Widrow training photographs above. Thus, each training photograph was made into 42 training patterns. The 42 training patterns for each person were stored in an individual memory folder and were labeled with the name of the corresponding person. There were 44 memory folders storing the training patterns for the 44 persons. In total we had a database of 1848 patterns.

The memory system was prompted with a set of 144 photographs from all 144 individuals. For the 44 specific people whose photographs were used for training, the prompting photographs for them were selected that were different from their training photographs.

By testing the prompting patterns with the trained autoassociative network, hits were obtained with 44 of the 144 prompting photographs. Figure 11.15 shows the distribution of minimum percent MSE obtained for each optimized prompting photograph. Persons 0–43 were trained into the autoassociative neural network; the remaining 100 were not.

Each of the prompting photographs was converted to many prompting patterns by varying rotation, translation, scaling, brightness, and contrast. By optimizing these adjustments, 44 hits were obtained.

The lower MSE's were obtained for the 44 trained-in persons. The high MSE's were obtained for the remaining 100 non-trained-in persons.

The 44 prompting photographs that generated hits corresponded exactly to the persons whose photographs were trained in. The remaining 100 photographs produced no hits. The hit photographs were compared on a pixel-by-pixel basis with the photographs stored in the memory folders. The selected patterns from the memory folders allowed identification of the 44 persons. The 44 prompt photographs of the trained-in persons were perfectly identified, and the remaining 100 prompt photographs were rejected. The 44 trained-in persons were perfectly identified, and those not trained in were perfectly identified as unknowns.

Another pattern recognition problem involved the use of photographs distributed by NIST for the Face Recognition Grand Challenge version 1. Two photographs of each of thousands of people were available. Single photographs of 75 individuals were used for training. For testing we selected 75 photographs of the same individuals. However each one of these photographs was other than the photographs used for training.

Additionally photographs of 300 persons not trained-in were also used for testing. In total there were 375 photographs used for testing.

As a result of this experiment, the cognitive memory system recognized and identified the 75 people that were trained-in without error, while 300 people not trained-in were rejected.

11.5 A Photographic Search Engine

Photographs are easily stored in computers. They can be downloaded from the Internet, from digital cameras, from digital scanners of analog photographs, etc. It is possible to have hundreds of thousands of photographs stored in a personal computer or other information appliance. The problem for the user is to be able to find a desired photograph, having no idea where this photograph may be stored. Below is described a methodology for searching for photographs stored on an information appliance, such as a desktop computer, laptop computer, handheld computer, digital media device, or mobile telephone.

If, for example, several hundred photographs of Paris were taken on a tour of Paris, they could be retrieved if they were placed in a directory named Paris, 2009. When seeking the continents of this directory, the user would ask for Paris, 2009. But if these photographs were simply recorded in the computer without any notation of Paris or any other labeling, they would be mixed with thousands of other photographs and difficult to retrieve.

It is common for users to store photographs without labeling. A photograph of a group of ten people would require some effort to locate and identify each person and list their names. If this were done, the photograph could be placed in a directory labeled with the list of names. Then retrieval would be easy, based on one or more of the names. But most people do not label each photograph when they are loading them into the computer. A search engine for photographs based on cognitive memory could retrieve unlabeled photographs containing multiple people, places, and things of interest. A conventional search would be based on keywords, which would not always be available.

On the other hand, a search engine for photographs can make a search based on a known photograph. For example, a given photograph of a person could be used to initiate a search for all photographs in the computer where that person appears. Photographs of two people could be used to initiate a search for all photographs containing both people, etc.

Referring to Fig. 11.16, a personal computer is shown with inputs from a digital camera, a photograph scanner, or an Internet connection. By these means, photographs can be downloaded into the computer. Photographs that can be used for search initiation can also be downloaded by these means. The personal computer is shown connected to a hardware neural network. Prompt patterns will be sent from the computer to the neural network. The neural network output patterns will be sent back to the computer through the connection between them. The entire photographic storage and retrieval system utilizes cognitive memory.

The computer may have a large number of photographs, perhaps hundreds of thousands, in bulk storage, without labels. To initiate a search, a known photograph must be available. This is the photograph of a person that the user wishes to search for among the photographs in bulk storage. From the photograph, many training patterns of the person's face can be generated by rotation, scaling, translation, and variation in brightness and contrast. These patterns are stored in a memory folder of the first memory segment and they comprise the database. These patterns are trained into an autoassociative neural network. This network and its training are implemented via computer software. The weights of the low-resolution and high-resolution neural networks that result from the training are now transmitted to the hardware neural network. Each photograph in bulk storage can now be very rapidly searched for the person of interest.

Each photograph drawn from bulk storage provides query patterns for the cognitive memory. These photographs prompt the cognitive memory. If prompting causes a hit, then the hit photograph contains the person of interest. In accordance with the configuration of Fig. 11.16, the training is done with software and the resulting weights are copied into many hardware neural networks, allowing simultaneous evaluation of many photographs at once.

Before each photograph can be analyzed in detail, the faces must be located within the photograph. This is accomplished by window scanning prompt patterns with a low-resolution system, as was done in Fig. 11.3. The low-resolution facial prompt pattern would be, for example, 20×20 pixels. In contrast, the high-resolution process would be, for example, 50×50 pixels. The weights for the low-resolution neural network are downloaded from the computer. The training to obtain these weights is performed off-line by the computer. The facial patterns used to train the low-resolution network were obtained from photographs from an arbitrary selection of individuals. This low-resolution scanning will find people's faces without identifying the people.

A search for the entire photograph by the low-resolution system will locate all of the faces present. For each face, the coordinates of the center of the hit window are recorded, noted along the horizontal and vertical axes in pixel counts. The photograph will be labeled in terms of the X–Y positions of all the faces present. Once this is done, high-resolution analysis can be started for each of the noted facial positions. The process is illustrated in Figs. 11.3, 11.4, 11.5, 11.6, 11.7, 11.8 and 11.9. If there is a hit, the photograph will be returned to bulk storage, now labeled with the coordinates and the names of the persons in the photograph. The hit photograph will then be delivered to the computer user. If there is no hit, the photograph is returned to bulk storage. The process continues with the next photograph from bulk storage, etc. (Fig. 11.10).

Most information appliances capable of sending and receiving emails have software address books containing contact information for people in communication with the user. This contact information would include the name of the person, phone numbers, e-mail addresses etc. It could also contain a photograph of the person.

This photograph can be used to generate patterns to train a high-resolution neural network. This can be done by the user's computer, and the resulting weights can be stored together with the photograph and other contact information in the address book.

If the person of interest is in the address book, then the neural network weights for this person can be determined, loaded into the low-resolution and high-resolution neural networks, and used to analyze the photographs in bulk storage. The speed obtained by the hardware neural network is required to do the search quickly. The weights of the high-resolution neural network can be determined for each individual and recorded in the address book while the computer is idle or not being used for another task. Also during idle time, the computer could go through one photograph at a time in bulk storage, using the low-resolution system to locate the faces, and using the high-resolution system to identify the faces as a named person in the address book or as an unknown person. The photograph can be labeled with names of the identified

Fig. 11.4 First face detected with the trained low-resolution network

people, and the locations of the known and unknown faces. The photographs in bulk storage will, over time, have labels and will be able to be recalled simply by naming the query person or persons. Thus this system can be used to label photographs that were not initially labeled, but were simply recorded in bulk storage.

The user will be able to retrieve photographs from bulk storage containing images of persons in the address book, asking for them by name. The computer can simply retrieve all photographs labeled with that person's name. New pictures of a person who may not be in the address book can also be used to initiate a search. Retrieval will take longer since a neural network will need to be trained for this person. A first search will be made to see if this person is already in the address book or not, and then search through bulk storage for relevant photographs will need to take place. High-resolution analysis will be done with the faces that have previously been detected, but identified as unknown. Where there is a hit, the photograph will be labeled and delivered to the user. It should be noted that photographs with identified people that are in the computer but not in the address book can also be used just like those in the address book.

Interesting analysis can be done with the facial images of people identified as unknown in the photographs in bulk storage. A second "address book" could be made from all of these images. The photographs in bulk storage could be numbered. The second address book would not have names attached to the unknown images, but would have the facial image number and the photograph number attached. An unknown facial image in one of the photographs can be connected to unknown images in other photographs. This would be "connecting the dots," finding all photographs with the same unknown person.

When the computer is idle, a second address book of unknown facial images can be formed. The search procedure is exactly the same as it was for the photographs

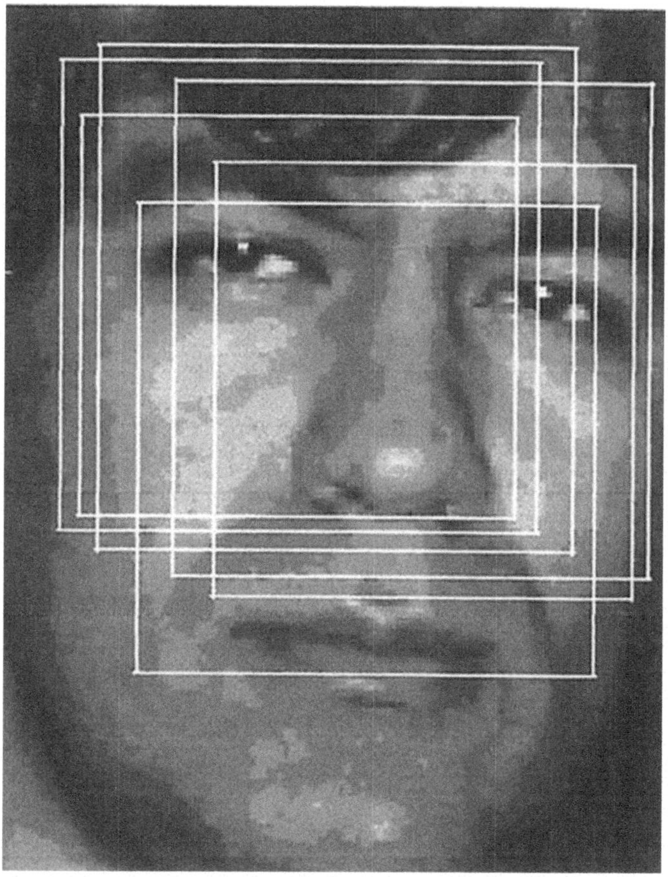

Fig. 11.5 Search process with the trained high-resolution network given the first detected face

of the first address book. The search will certainly find the photograph from which the unknown photograph was taken. But the search will also find other photographs containing images of the same person, giving the numbers of pictures in bulk storage and the coordinates of the person's image in those photographs.

Another type of search for both documents and photographs stored in the computer can be made in response to a query that has both text and the photograph. With text having keywords, the search algorithm can call for all documents that contain the keywords and have photographs that in turn correspond to the prompt photograph. The photographic search can be done with the methods of Fig. 11.16. The result of the search is a document containing a photograph of interest.

Yet another type of search, this time for photographs, can be made with a query containing a photograph and text having keywords. The text can be used to initiate a search through all directories labeled with the keywords, and through all directories containing documents containing the keywords. The contents of all these directories

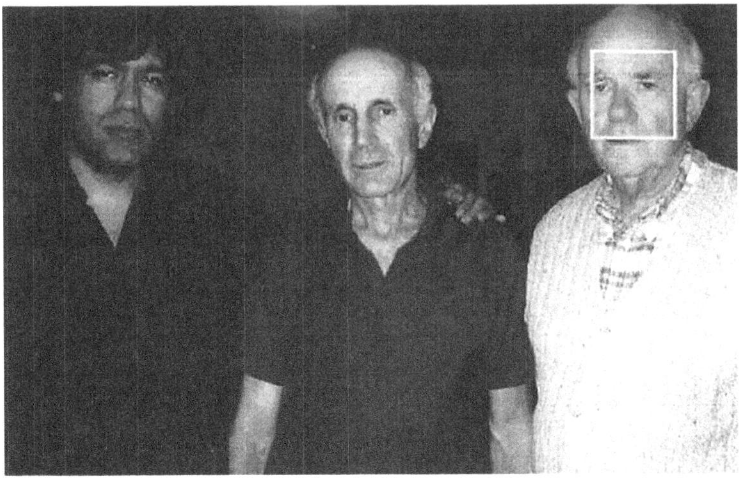

Fig. 11.6 Second face detected

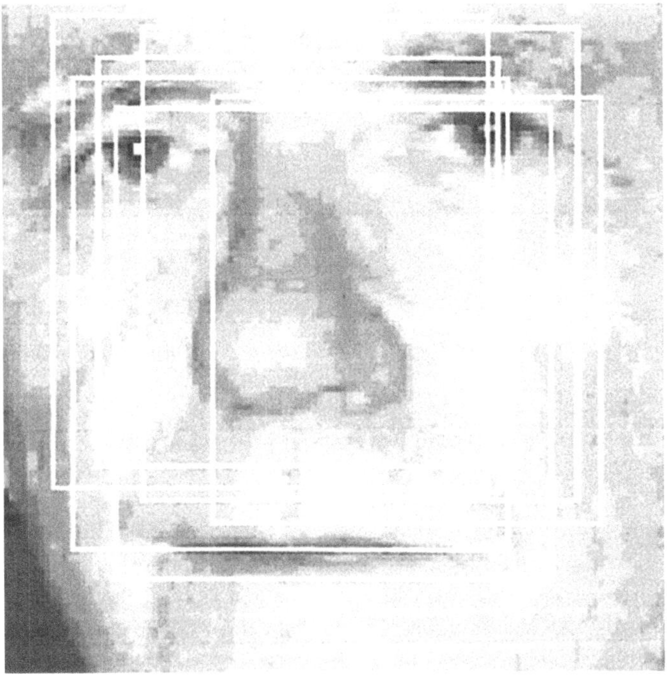

Fig. 11.7 Search process with high-resolution network for the second detected face

Fig. 11.8 Third face detected

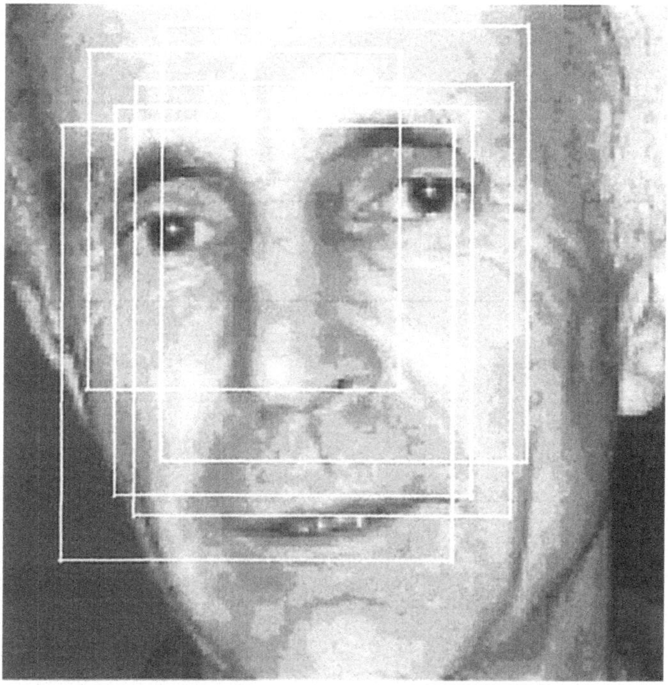

Fig. 11.9 Search process with high-resolution network for the third detected face

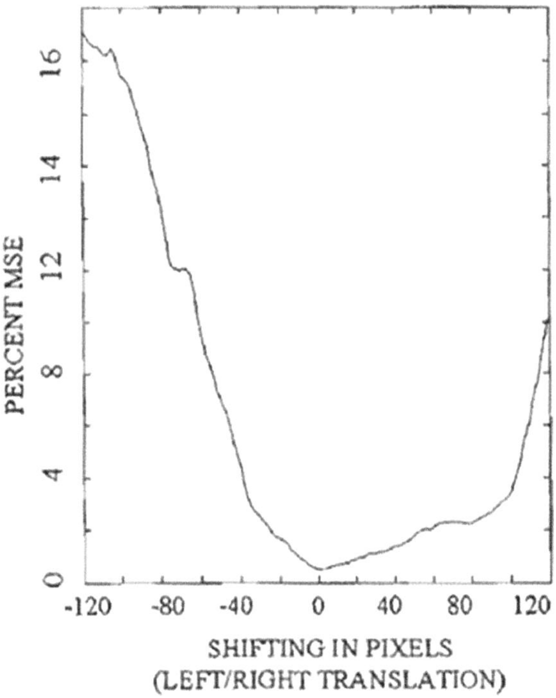

Fig. 11.10 Effects on mean-square-error due to left/right translation from the optimal position

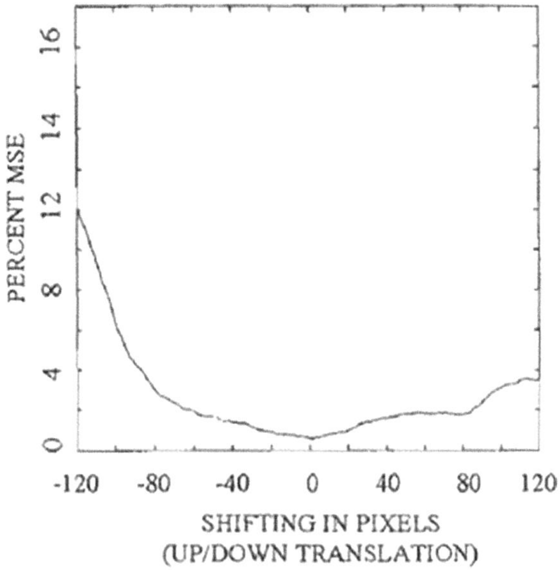

Fig. 11.11 Effects on mean-square-error due to up/down translation from the optimal position

Fig. 11.12 Effects on mean-square-error due to rotation from the optimal orientation

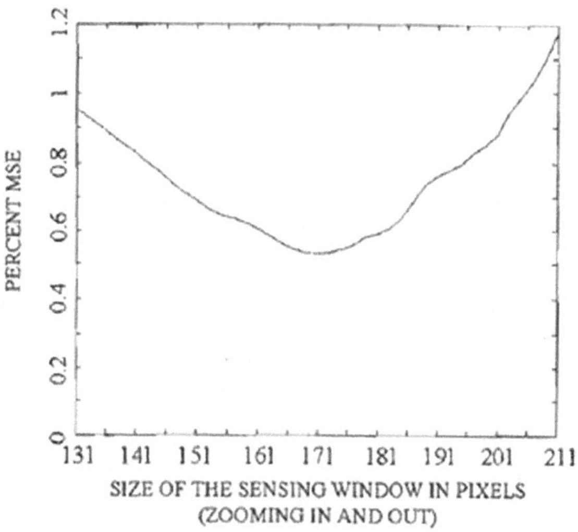

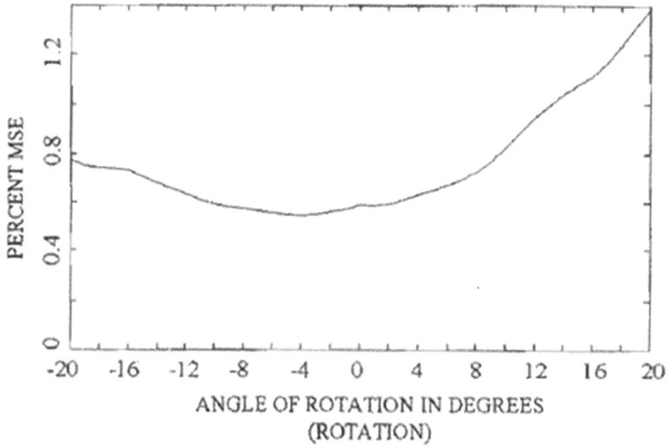

Fig. 11.13 Effects on mean-square-error due to zooming from the optimal scale

can then be searched to find photographs that correspond to the query photograph. The results of the search will be photographs corresponding to the query photograph and relating to the keywords. An example might be "Paris" as a keyword, and a photograph of the Eiffel Tower as the prompt photograph. The system would deliver photographs of the Eiffel Tower that are in various directories and that correlate with the prompt photograph.

An extension of the photographic search could include searching for video files. If the individual frames of a video clip are treated as still photographs, all of the above processes can also be applied to search for files containing this video, in response to a query containing a query photograph, or text and a query photograph.

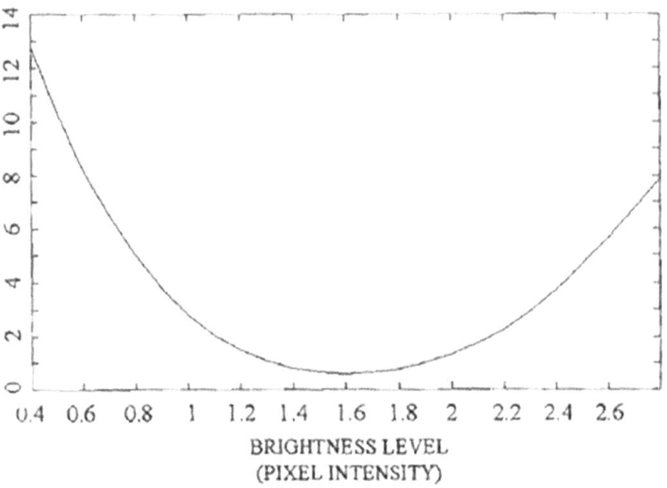

Fig. 11.14 Effects on mean-square-error due to variation of image brightness from the optimal luminance

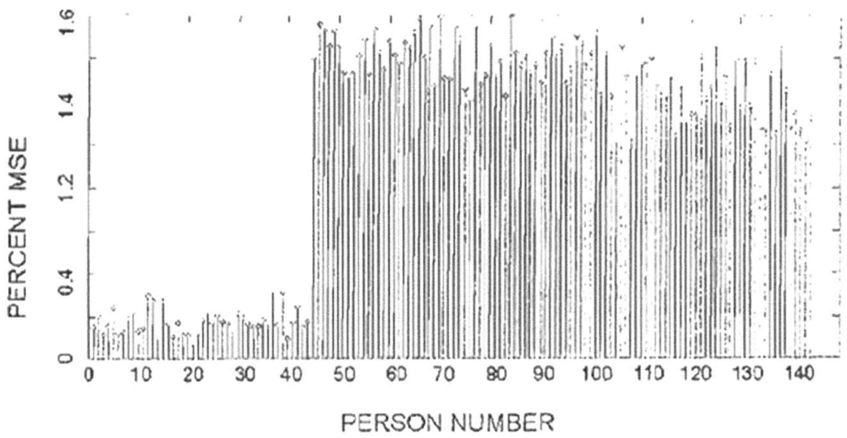

Fig. 11.15 Distribution of minimum percent MSE obtained for a group of 144 individuals

The search engine of Figs. 11.11, 11.12, 11.13, 11.14, 11.15 and 11.16, used for searching a personal computer is unique. This search engine is very fast and accurate, based on the cognitive memory concept.

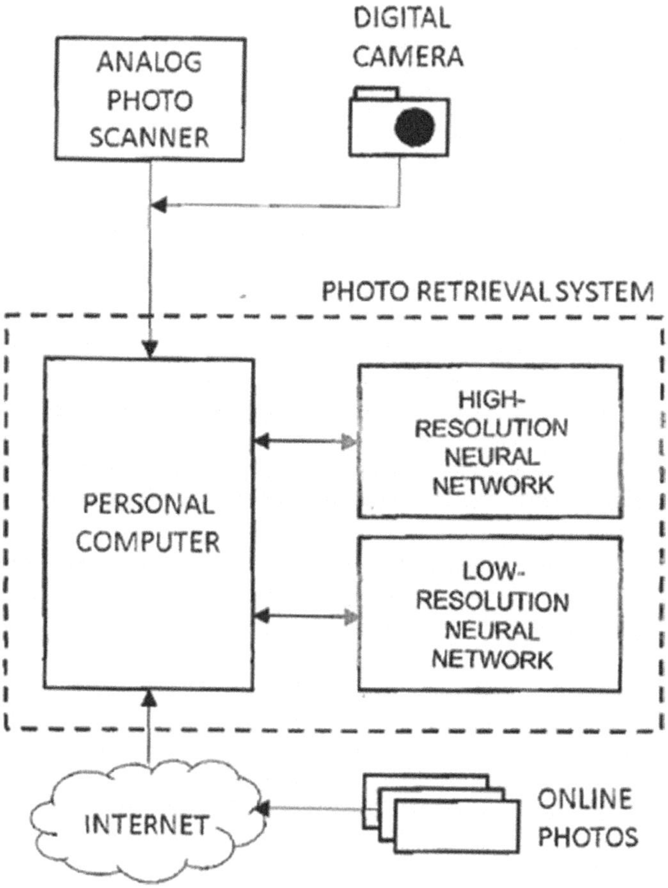

Fig. 11.16 A photographic search engine for a personal computer

11.6 Summary

Cognitive memory can be utilized for human face recognition. The methodology of a human face recognition process can be summarized by the following steps:

(1) Individual photographs or other images of a multitude of persons and their respective identities are stored in a face recognition system.
(2) A multitude of images may be created from individual photographs of identified persons by any combination of translation, rotation, scaling, brightness adjustment, or contrast adjustment, and stored in the face recognition system with the same respective identities.
(3) The faces in the photographs or other images from steps (1) and (2) are located and trained into one or more autoassociative neural networks, which are part of the face recognition system.

(4) A query photograph, a photograph or image presented to the face recognition system for identification that may contain one or more faces to be identified, is presented to the face recognition system.
(5) All facial images that may be present in the query photograph are detected and located.
(6) All portions of the query photograph containing located facial images may be pre-processed by any combination of translation, rotation, scaling, brightness adjustment, or contrast adjustment to create prompt patterns for the autoassociative neural network.
(7) The various prompt patterns are applied as inputs to the trained autoassociative neural network and, if the error in the difference between the prompt pattern input and the resulting autoassociative neural network output pattern is below a pre-set threshold, there is a "hit," and the prompt pattern is considered successful. If there is no hit, no faces are identified.
(8) If there is a hit, the successful prompt pattern is compared with all stored original facial images and all facial images that may have been created from them to find the best match.
(9) The best matching photographs or images and their identities will be delivered as the output of the face recognition process.

Appendix

Use google (or your favorite internet search engine) to explore:

- Wikipedia face recognition
- Noisy photo
- Cluttered photo
- Recognition of a face among many
- Image preprocessing, translation
- Image preprocessing, rotationing, scaling.

Chapter 12
Cognitive Memory for Navigation

Abstract An aerial photograph of Simi Valley, California, was obtained. This snapshot was taken by an airplane flying at a known altitude and archived. This photo was available to a second airplane whose position is to be determined. Equipped with a downward looking telescope, the second airplane flies at the same altitude as the first airplane. The problem addressed in this chapter is that of locating the position of the second airplane with respect to the original aerial photo by comparing the telescope image with the aerial photo. The aerial photo is like a map. The second airplane is able to locate its position on the map. This method, based on cognitive memory, can be exercised repeatedly, producing a track of the second airplane on the map. Comparison of the telescopic image with the map image involves translation and rotation with the goal of seeking a 'hit'. If the second airplane was flying at an altitude other than that of the first airplane, a scale change would be required.

In everyday life, memory plays a major role in navigation. For example, in a large house with many rooms, a person can go from one room to another at will without giving it much thought. Having lived in the house, information learned and stored in memory provides guidance. Suppose you were at home and decided to go outside and take a walk to buy some ice cream. Walking along, visual cues are noted along the path to provide guidance.

After a long day at work, driving home is automatic. Visual cues in the street correlate with memory allowing one to know where to turn in going from one street to another. Driving home, generally GPS is not used. Visual cues are sufficient to navigate. Driving home at night is more difficult. Visual cues are more challenging to find in the low light conditions. Visual cues can be used for aircraft navigation as will be demonstrated herein.

B. Widrow and E. P. Katz, *Cognitive Memory*, Springer Series on Bio- and Neurosystems 17, https://doi.org/10.1007/978-3-031-80939-2_12

12.1 Aircraft Navigation

A previously recorded aerial photo of a city or of a countryside with objects of interest in view is assumed to have been obtained. Now an airplane flies at constant altitude over the land area seen in the aerial photo. A telescopic camera mounted in the "belly" of the airplane takes straight-down digital snapshots of the ground below. The objective is to be able to locate the X–Y position of the airplane relative to the aerial photo and to determine its direction of flight at each instant when a snapshot is taken. The question is how to do this in real time with an on board computer, while flying over complicated territory such as a city.

12.2 Simi Valley

Simi Valley is located in Ventura County, California, about 40 miles from Los Angeles. An aerial photo of Simi Valley was downloaded from a website (Fig. 12.1). For purposes of recognition, the photo has been "diced," cut up into small squares. The position of the photo within each square is a pattern that will be fed to a cognitive memory for training. Each diced square contains a small 81 × 81 pixel image. The entire photo contains 2700 × 2700 pixels. The total number of 81 × 8 1 images is 918. Only 170 small images of the total photograph are shown in Fig. 12.1.

The small images are converted to black and white and are reduced to 25 × 25 pixel patterns using interpolation and resampling, in order to reduce the number of dimensions of each small image. The images are converted to vectors with the intensity of each pixel being a component of the corresponding vector. Each component ranges from − 0.9 to + 0.9 after quantization. Each small image is represented by a 25 × 25 = 625 component vector.

All 918 small images are trained into the autoassociative neural network of a single memory segment. Initial weights are set randomly. There are 250 neurons in the first layer, each with 625 inputs. The first layer has a total of 625 × 250 = 156,250 weights. The second layer had 420 neurons, each with 250 inputs. The total number of weights of the second layer was 420 × 250 = 105,000. The third layer had 625 neurons, each with 420 inputs. The total number of weights of the third layer was 625 × 420 = 262,500. For the entire network, the total number of weights was 523,750. The total number of neurons was 1295. Training was performed using the back propagation algorithm. The 918 small images, pre-processed to 25 × 25 black and white images, are stored in the segment's memory folder. In the memory, each image is identified with the X–Y coordinates of its center. The images now represented as vectors are trained into the neural network.

The objective is to be able to identify the location of an airplane flying over Simi Valley relative to the aerial photo of Simi Valley. The telescope in the belly of the airplane obtains a view of the ground. A camera on board the aircraft takes snapshots of the telescopic image. Figure 12.2 shows a representation of an airplane flying over

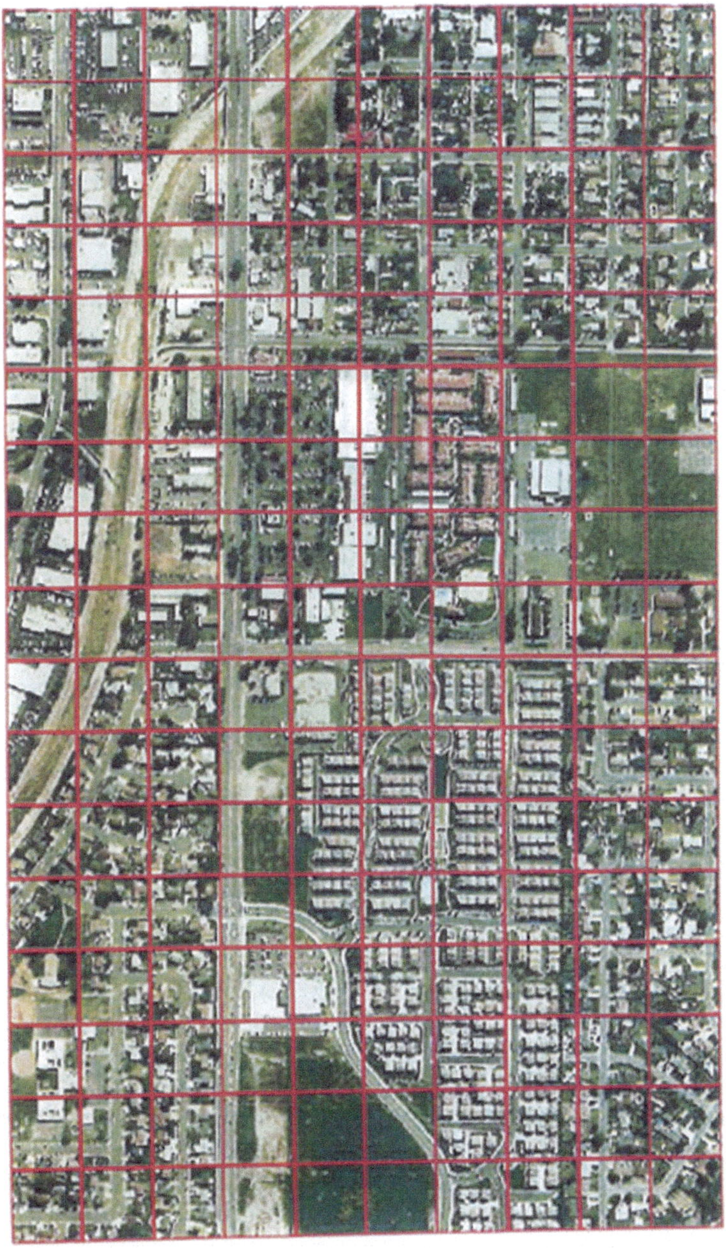

Fig. 12.1 Diced aerial photograph of Simi Valley, CA

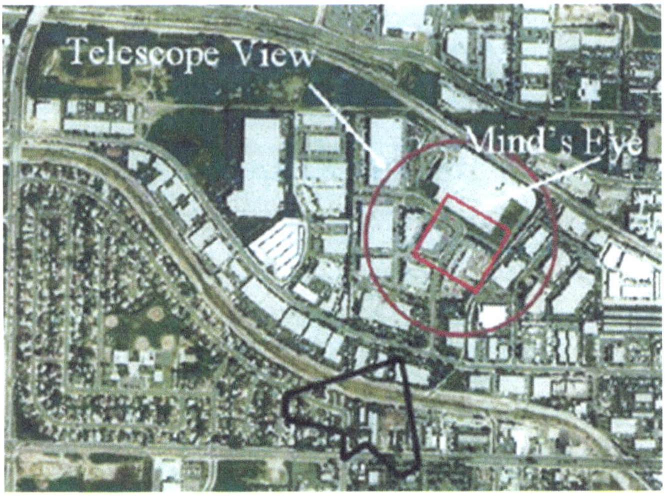

Fig. 12.2 Telescopic view of airplane and mind's eye image

Simi Valley. The circular telescopic view is also seen. The center of the circle is the current position of the airplane.

Within the telescopic view, a square window is shown. The size of that window is designed to match the size of the original grid over the aerial photo of Simi Valley. The airplane is assumed to be flying at constant altitude.

The telescopic image is fed into the onboard computer. The square window can be moved in software via left or right translation, and it can be rotated. The square window is called the "mind's eye," as if a person were looking through the telescope and varying the viewing direction. The image within the square window is the prompt for the cognitive memory.

Given a snapshot, the square window can scan over it with various left–right translations and a variety of rotations. When a combination of translation and rotation causes a primary "hit," the multiplexer (see Fig. 3.2) emits all 918 small images and their X–Y location in pixels in a sequence. A secondary hit (see Fig. 3.6), identifies the small image in the aerial photo and its X–Y center, corresponding to the image within the onboard square window. Knowing the amount of translation and rotation that was needed to make the primary hit, the position of the airplane and its heading can be computed relative to the "hit" small image of the aerial photo. Thus the position and heading of the airplane can be located on the original aerial photo.

The airplane was flying at a constant altitude in a circular arc over Simi Valley. Snapshots of the telescopic image were taken at uniform time intervals. The position and heading of the airplane were determined at the time of each snapshot.

Figure 12.3 is an aerial photo of Simi Valley. The arrows delineate the flight path. The tail of each arrow indicates the position of the airplane at the time of each snapshot. The direction of each arrow indicates the heading of the airplane at

Fig. 12.3 Circular airplane track over Simi Valley, CA

the particular time of the snapshot. The circular arc is clearly seen. Note that the algorithm did not know that the path was circular.

If the proper size of the square window (the mind's eye) were not known, then the square window would need to be scaled. This would be in addition to translation and rotation. All of these are needed to search for a hit. More computation would be necessary, but this could be feasible.

In Fig. 12.3, the positions of the arrows were located within one pixel in X and Y. When scaled for the aerial photo, one pixel represents approximately a 10 foot square on the ground. The arrows are tangent to a circular arc. Their directions are accurate to within about $\pm 1^\circ$. The airplane is tracked relative to the original aerial photo. If GPS is used, the airplane would have been tracked relative to a geographic reference frame.

12.3 A Summary of the Method

Before flight

- Obtain an aerial photo of the area of interest.
- Dice the aerial photo image.
- Feed the diced images to a cognitive memory system. Record them in a single memory folder of segment 1, and record their X–Y locations, in pixels, on the aerial photo.
- Train the autoassociative neural network with all the diced images.

During flight

- Copy the weights of the converged autoassociative network into an identical network implemented on an onboard computer.
- Record the current circular telescopic image of the ground taken from a snapshot.
- Place a square "mind's eye" window within the circular telescopic image. Initially center the mind's eye at the center of the circular telescopic view.
- Vary the X–Y position and angular orientation of the mind's eye view relative to the telescopic view. A hit will occur when the variation finds a hit image that matches one of the images stored in memory.
- Matching pattern to pattern, and exhaustive search of the memory folders will find the pattern identical to the hit pattern. The X–Y position of the hit pattern in the memory folder is recorded. The X–Y position and orientation of the airplane is determined from this, and the optimal X–Y displacement and angular displacement of the mind's eye to make the hit.
- The size of the square mind's eye should be of the same dimensions as the original diced images. If this is not so, then the scale of the mind's eye would be varied along with the X–Y parameters and the angular orientation in order to achieve a hit.

This process locates the instantaneous position and orientation of the airplane on the original aerial photo. This is basically navigating while looking downward on a road map.

12.4 MSE Sensitivity to X–Y Position and Angle Orientation

The autoassociative neural network of the cognitive memory is trained with the 918 (25 $\times$ 25 pixel) diced images. When the mind's eye is oriented to make a hit, the unique correct image of 918 possibilities should be selected. It is important in making the hit that the optimization of the positional and angular parameters be done precisely to minimize the MSE. The question is, "*What would happen to the MSE if the choice of positions and angle are slightly offset from optimal?*" In other words, how sharp is the discrimination of the autoassociative network?

Figure 12.4 shows the MSE as a function of positions and angles. The sharp dip in MSE results when the parameters are optimized. In this plot, the Y-parameter is fixed at its optimal value. The X-parameter is varied, and the angular parameter is varied, starting from optimal. Slight deviations from optimal can cause large increases in the MSE.

In Fig. 12.5, conditions are the same as with Fig. 12.4, except that the Y-parameter is set at one pixel less than the optimal. The depth of the dip is substantially reduced, even when the X-parameters and the angle parameter are re-optimized. A hit may or may not be recorded, depending on the setting of the threshold. An error of one pixel

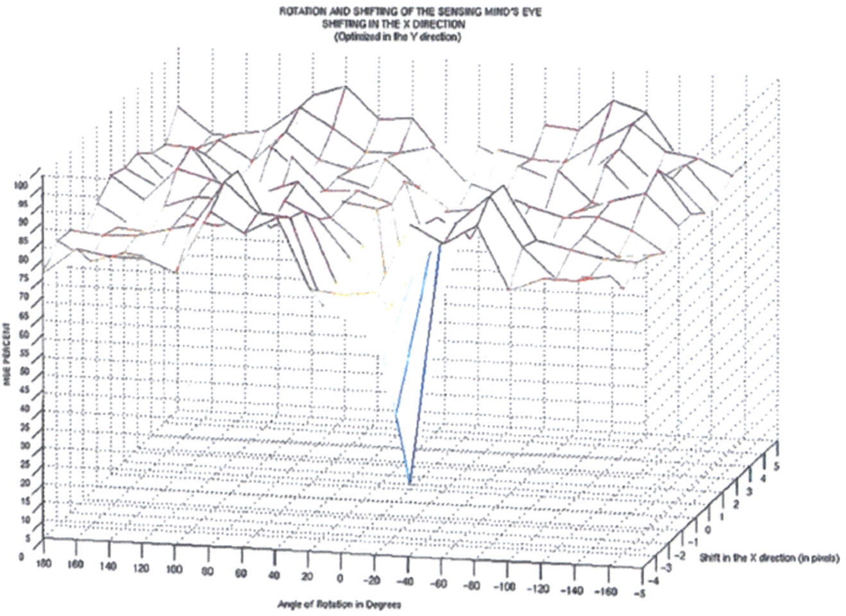

Fig. 12.4 Rotation and translation of the mind's eye (optimized in the Y direction)

has a large effect on the MSE. The conclusion is that the autoassociative network will allow sharp discrimination in the position and angle parameters.

12.5 Another Example

Figure 12.6 shows an aerial photo of Washington, DC. This photo is diced in a similar manner resulting in 900 images which are used to train into the autoassociative network.

An airplane flying in a straight line path over Washington took snapshots over uniform time intervals. The position and heading of the airplane at the time of each snapshot is noted. Figure 12.7 shows a series of white arrows, each corresponding to a snapshot. The arrows indicate position and heading angle. Placing the arrows onto the aerial photo, a straight line can be observed. The cognitive memory did not know the flight path. The result shown in the black and white Fig. 12.7 resulted from the algorithm.

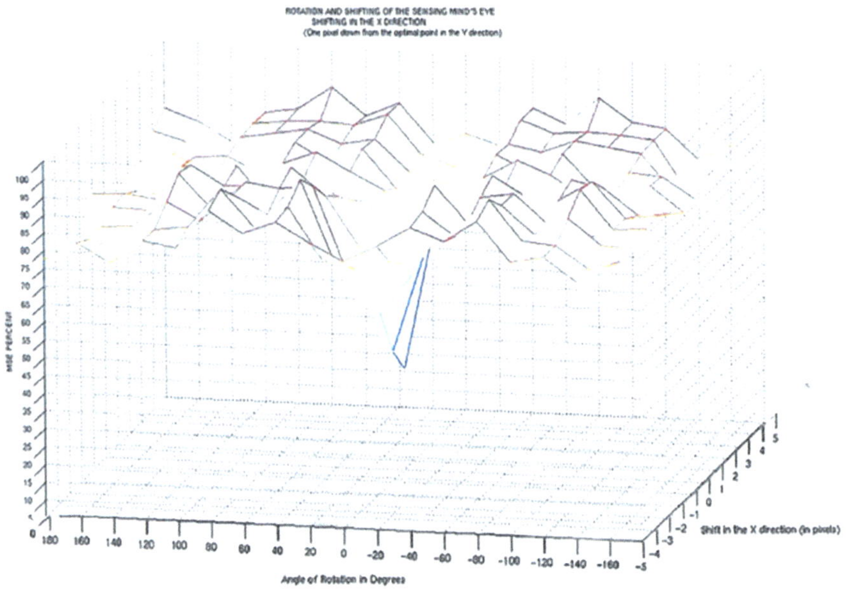

Fig. 12.5 Rotation and translation of the mind's eye (One pixel down from optimal point in Y direction)

12.6 Summary

A map-like aerial photo of Simi Valley is obtained. A square grid is drawn on the map. Hundreds of squares are formed. The image within each square is entered into a cognitive memory as its database, each image labeled with its position. An airplane flying over Simi Valley is equipped with a downward looking telescope. Within the circular field of view of the telescope, a square is placed that could be varied by translation, rotation, and scale, creating hundreds of images for every view through the telescope. These images become prompts to the cognitive memory. When making a "hit," the square image in memory and its location is outputted. Using this and noting the translation, rotation and scale needed for the hit, the position, heading, and altitude of the airplane can be computed, all with respect to the original aerial photo.

Fig. 12.6 Diced aerial photograph of Washington DC

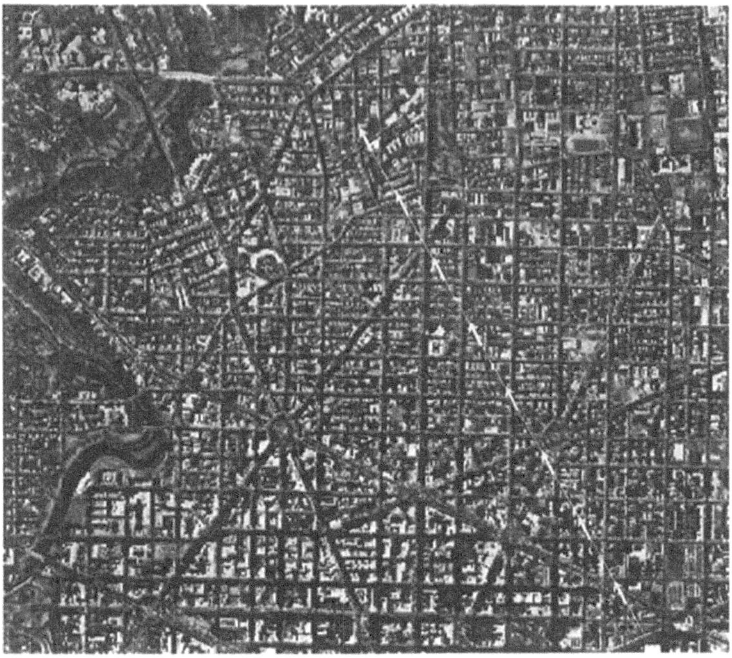

Fig. 12.7 Straight-line airplane track over Washington

Appendix

Use google (or your favorite internet search engine) to explore:

- Simi Valley, California
- Wikipedia, Simi Valley
- Wikipedia, aerial photography
- Aerial photo-map-based navigation.

Part IV
Human Cognitive Memory

We all have had experience living with a human memory. The cognitive memory described in previous chapters was designed based on this experience. It is the purpose of Part IV to further explore human connection between human experience and the concept of a cognitive memory.

The working of a mechanistic cognitive memory can be understood by its creators, an engineer or a computer scientist. The living memory has workings that are not understood. One can only speculate about this. Part IV speculates, but the speculation is based on living experience.

Chapter 13
Cognitive Memory for Seeing and Hearing

Abstract Human memory has both short-term and long-term components. To retrieve information from long-term memory, the memory must be prompted. To retrieve information from short-term memory, prompting is also necessary. Storage in short-term memory is temporary, and described items are chosen based on a perception of time, i.e., a sense of when an event happened. Items from short-term memory can be prompts for long-term memory. When executing a procedure that is stored in long-term memory, the current step of the procedure recorded in short-term memory prompts the long-term memory to provide the next step. This is known as "memory inertia."

13.1 Cognitive Memory for Pattern Recognition

The proposed cognitive memory system performs pattern recognition but differs significantly from the usual pattern recognition systems [1, 2]. Most adaptive pattern recognition systems learn to classify patterns by adapting to a set of training patterns. Once trained, the training patterns are discarded. When new input patterns are applied to the trained classifier, these patterns are classified in accordance with the training experience [1].

Long-term cognitive memory, on the other hand, stores the training patterns (visual, auditory, etc.) in folders and recalls the entire contents of a folder when prompted by a new input pattern that may match only one set of patterns in that particular folder. If the patterns in the folder are identified by their meanings or classes, then the new input pattern gains meaning or class by induction or association.

A good example is the analysis or diagnosis of mammograms. Systems have been in development over many years for the automatic analysis of mammograms, spotting lesions that may be cancerous [1, 3, 4]. These systems are based on neural networks and are trained or designed to classify mammographic X-ray images. They have a record displaying some success. New unknown mammogram images have been applied to the trained system and have, in most cases, been classified correctly.

B. Widrow and E. P. Katz, *Cognitive Memory*, Springer Series on Bio- and Neurosystems 17, https://doi.org/10.1007/978-3-031-80939-2_13

A cognitive memory could also be applied to this problem. Folders for thousands of patients could be established, each containing a series of mammograms taken from a particular patient over a number of years. The patient's history giving the age, weight, height, blood work, biopsy, the treatment given, and the outcome would all be recorded in the folder. When a new patient comes into the system, the new mammogram prompts a search and, adjusting the tolerance or closeness of the image matching, hits take place for a number of folders. The physician then has several previous cases that are similar to the present one. From this experience, a treatment plan can be formulated.

The way a physician currently addresses this problem may be similar to the cognitive memory approach. The physician has many years of experience reading mammograms and treating patients [5]. When a new problem arises, the physician thinks of previous similar cases. What was done? What was the outcome? "What do I see?" and "What do I know about that?".

13.2 Learning

In an animal brain, a synapse is the coupling that connects one neuron to another. In an artificial neural network, neurons are coupled from one to another by what are called weights. In both the living and artificial neural networks, the strength of the coupling is variable. The synaptic weight or coupling varies while learning. While learning takes place, weights can increase or decrease in strength and their values can fluctuate slowly. All of the weight values together represent the memory of the neural network. Learned responses are stored in the weights. Training neural networks is one form of learning.

Most profoundly, learning is the storage of images or data in long-term memory. This is what happens within a cognitive memory. Sensory images that could be visual, auditory, etc. that are "interesting" are stored in permanent memory. This is the storehouse of the knowledge database. Learning is the process of recording images and data into long-term memory.

13.3 What is "Interesting"?

What is interesting will vary from person to person. Recording in long-term memory takes place when the "interesting" switch represented in Fig. 13.1 is closed and allows input signals streaming from visual, auditory, etc., sensors to pass through to the memory input line.

With long-term memory, prompting signals come from the same sensors as those from which the original data was recorded. Prompting signals are tested by the same "interesting" switch and if accepted, they will be transmitted to the prompt input

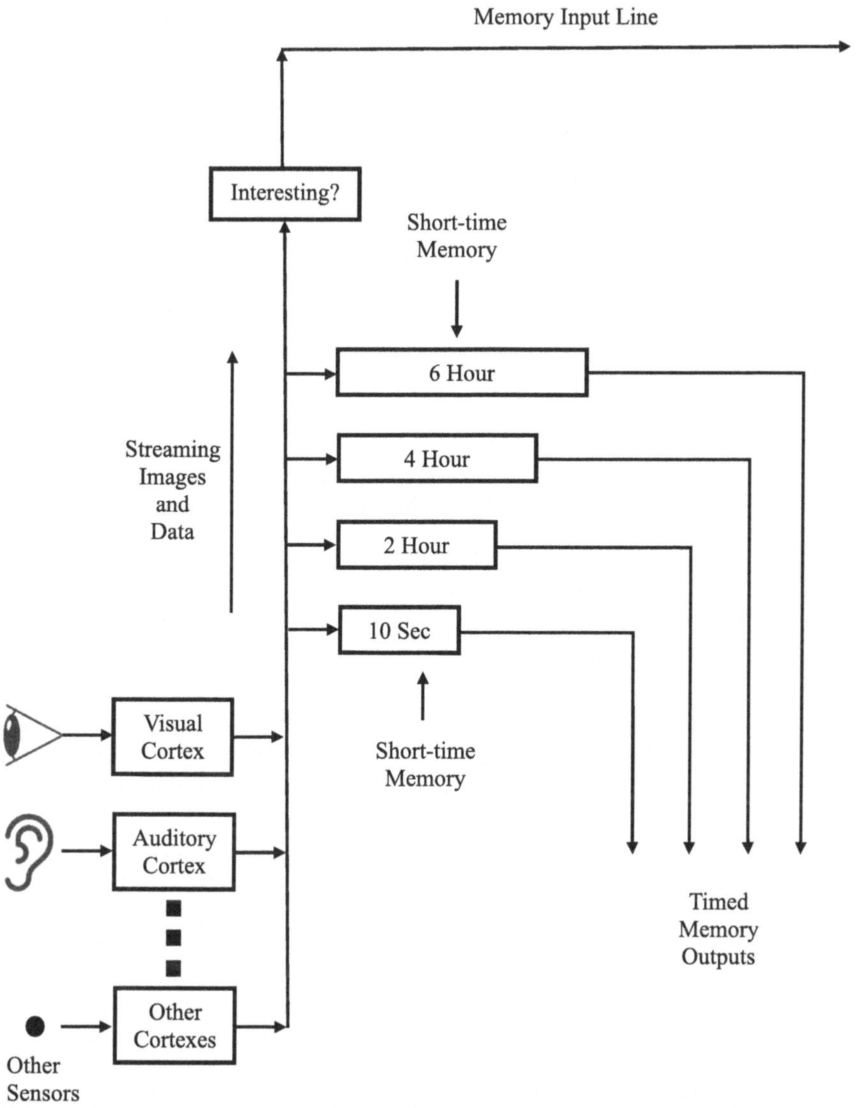

Fig. 13.1 Recording and retrieval for long-term and short-term input images and data

line. Sensory signals must pass through the "interesting" switch for both memory recording and for prompting and memory retrieval.

The "interesting" switch could be activated by a loud sound, a bright light, spectacular color, etc. The content of the sensory inputs could also activate the "interesting" switch if these inputs, when tested against the contents of the entire memory, would cause a hit somewhere. With a hit, the new incoming signals related to something seen before and are therefore interesting. The "interesting" switch could also be activated by volition, a desire to record new images even if they do not meet the other criteria.

13.4 Long-Term Memory/Short-Term Memory

The input sensory images and data streaming into the memory are tested by the "interesting" switch for recording or retrieval, for dealing with long-term memory. Short-term memory does not need testing. Everything is recorded.

Short-term memory units are indicated in the diagram of Fig. 13.1. Without the "interesting" testing, sensory inputs are applied directly to short-term memory units. Signals are streamed constantly into these units. They behave like shift registers where new inputs are accepted and the oldest signals are discarded. The memory amount for these units is limited. The shortest-term unit has a memory time of about 10 s. Other units have longer memory times.

The shortest memory contains all inputs over the past 10 s so that the human can see or hear what has transpired over the past 10 s. Likewise for memory units of 2, 4, 6 h, etc., that have recordings of the past 2, 4, 6 h, etc. (These recording times are hypothetical.) For retrieval, the human user selects the memory unit based on an approximate perception of time.

The shortest memory unit is selected by time of occurrence. Specific items are further selected within the time frame by prompting based on content. Short-term memory retrieval is based on time perception and on content. The finite recording times of the short-term units help to make searching easier.

If someone were to give you a 10-digit telephone number and asks you to make the call on a telephone across the room, you will be repeating this number over again in order to not forget it. It is estimated that the shortest of the short-term memory unit holds only about 10 s of data. The memory times of the other memory units seen in Fig. 13.1 are only estimates, not based on solid data.

13.5 The "Inertia Memory"

Have you ever gone from one room of your house to another room to do something, and upon arrival in the second room, you have forgotten why you went there? This is a failure of the shortest short-time memory unit, the 10 s unit.

When you are in the middle of a project, the 10 s memory unit keeps you going, remembering what you have just done serves as a prompt to what you need to do next.

When attending a concert featuring a piano concerto, you notice that the soloist "knows the work solid." The music is stored in long-term memory. There are no music sheets on the piano. The sound of the music just played by the orchestra and by the soloist acts as a prompt for the soloist to know what comes next.

The 10 s memory unit could be called the "inertia memory." By prompting the long-term memory, it maintains continuity and forward progress when executing a project or developing a thought..

13.6 A Story

Here is a Bernard Widrow story:

> I attended a concert of the San Francisco Symphony on April 7, 2007, that featured two works by Igor Stravinsky. The program notes told a Stravinsky story. At the age of 11, he was taken to a gala performance at Saint Petersburg's Mariinsky Theatre of Glinka's, "Ruslan and Ludmila." "Igor, look, there is Tchaikovsky," said his mother. He looked and saw a man with white hair, large shoulders, a corpulent back, and "this image has remained in the retina of my memory all my life."

Stravinsky was about 85 years old in 2007.

For recall from both long-term and short-term memory, a great deal of information will be present, most of which will be of no interest. Extracting the information of interest from the total recollection will be a cognitive task, but at least all of the sought information will be there. It is a matter of separating the wheat from the chaff.

13.7 The Identi-Kit

When police are interviewing an eyewitness to a crime, they often ask for a description of the suspect. If they had a pencil and paper, they might ask the witness to draw a sketch of the suspect. Doing these things is difficult or impossible for the eyewitness to achieve.

A very clever device called the Identi-Kit is now used to enable a more accurate description. The Identi-Kit serves as a prompt. It shows a face that is adjustable. Different features can be altered until the witness indicates that the likeness is a good fit for the remembered face from the scene of the crime. The image of the adjusted Identi-Kit is prompting the adjustments to be made. The answer is negative until the adjustments converge and provide an image that approximately matches the true image of the suspect's face, as well as can be remembered by the witness. The similarity of the converged Identi-Kit image to a photograph of the perpetrator is generally quite remarkable. The Identi-Kit image provides the police and the public with a resembling face for which to search.

13.8 Cognitive Memory with Auditory Inputs

Long-term and short-term memory for auditory signals is configured like that for visual images. Auditory and visual images are stored separately, but in the same folder. The means of long-term storage and retrieval for auditory images is very similar to that for visual images. The mechanisms for short-term storage and retrieval are the same as for visual and auditory images, signals, and data.

Long-term visual images are varied in the visual cortex by rotation translation scale, etc. so that many versions of the original image can be recorded. A single image might become thousands of images, all to be stored. Likewise, prompt patterns are varied so that many versions would be available for testing and searching for a hit. These variations take place in living systems in the visual cortex.

Auditory signals would also be varied for recording and prompting. They could be varied in amplitude. They would be stretched (distended) in time, or they could be compressed in time. Their spectral distributions could be distorted in many ways. Other forms of variation are possible. They could be filtered with nonlinear filters. The possibilities are endless. Experience would be the guide in exploring combinations of distortion that could exist in living systems, performed by the auditory cortex.

13.9 "Visual Cortex" Realization

Widrow and Winter in a 1988 paper [1] showed how neural networks could vary visual patterns by left–right and up-down translation, rotation, scale change, and combinations. They also showed how single neurons can perform logic functions, such as `and`, `or`, and `majority vote-taker`. Techniques such as these could be used by nature to implement the indicated functions in the visual cortex. Although not mentioned in that paper, it is clear that similar techniques could be used in the auditory cortex to provide variations.

13.9.1 Speculation About Seeing and Hearing

- Seeing involves processing and recording new visual images and making associations with pre-recorded images stored in memory.
- Vision and memory are intertwined.
- Hearing and understanding speech involve processing and recording new auditory images and making associations with pre-recorded auditory images stored in memory.
- Hearing and speech understanding, and memory are all intertwined.

13.9.2 Postscript: Generative AI

During the past decade, a novel form of AI that uses neural networks has emerged called *generative AI* which also has another form known as *generative adversarial networks* (GANS). Ian Goodfellow is the pioneer in this field [2, 3], with an astonishing number of applications of GANS which has cemented this into the soul of AI. This subject is mentioned here because of its strong similarity to that of the Identi-Kit technique.

There are so many applications of GANs that it is difficult to select one to illustrate the methodology. A clear and simple choice is that of creating visual images, given a piece of text as a prompt.

With GAN, there are two neural networks that are in some sense adversarial. In the text to image example, one network is trained to generate images from "scratch." These images are presented one-at-a-time as inputs to the second neural network. The two neural networks have been trained independently. The second network converts images to text and serves as a "sanity check." For each generated image, the second network's text output is checked against the original text. Where there is correspondence, the system outputs the approved generated image. Both neural networks are trained with large amounts of independent data culled from the internet.

The analogy to the Identi-Kit is clear. The Identi-Kit, and its operator, plays the role of the first neural network, generating many trial images. The witness selects the most acceptable image. The witness plays the role of the second neural network, selecting an image based on a remembrance of the face of the suspect at the crime scene. The witness does the "sanity check" using the Identi-Kit image to prompt his/her memory.

13.10 Summary

Sensory signals must pass through an "interesting" switch before being recorded in long-term memory. These sensory signals are recorded automatically by short-term memory, whether they are interesting or not. Short-term memory consists of many components, some storing data over the last 10 s and some storing data over the last 10 h. The memory components act like a computer shift register where new data enters and the oldest data is discarded such that data moves along the line. Short-time data is accessed by the perceived time when an event happened. Access is based on time, and with prompting on the content. Data stored in short-term memory can serve as prompts for long-term memory, if interesting.

Appendix

Use google (or your favorite internet search engine) to explore:

- Memory retrieval
- Memory storage
- Prompting long-term memory
- Prompting short-term memory
- Wikipedia, search for Ian Goodfellow.
- Wikipedia, search for generative AI.

References

1. Widrow, B., Winter, R.G.: Neural nets for adaptive filtering and adaptive pattern recognition. IEEE Comput. Mag. **21**(3), 25–37 (1988)
2. Goodfellow, I., Bengio, Y., Courville, A.: Deep Learning. MIT Press, Cambridge Mass (2016)
3. Goodfellow, I., Yoshua, R., Aaron, A.: Deep Learning. MIT Press, Cambridge, MA
4. 5 big predictions for AI in 2017. Will Knight, MIT Technology Review (2017)
5. Douglas, W., Carter, E.: The Year Creativity Exploded. MIT Technology Review, pp. 40–46 (2023)

Chapter 14
Cognitive Memory for Motor Control: Walking and Speaking Improvement with Reinforcement Learning

Abstract The body's muscles for walking, speaking, lifting, pushing, etc., are activated by neural control signals that originate in the motor cortex of the brain. In the motor cortex, large arrays of neural signals are generated to enable the desired muscle activity. Because of their complexity, it is unlikely that these control signals are generated in real-time. More likely, these control signals are drawn from memory after training for the respective tasks. That being the case, how is the cognitive memory trained, and once trained, how are the prompt signals generated to retrieve the control signals from memory? Training is done with reinforcement learning. Training is a slow, unsupervised process. When the muscle motion is satisfactory, the memory's neural networks are rewarded. When not satisfactory, the neural networks are punished. Prompting signals come from the physical environment. When walking, for example, the prompt signals for the next step come from sensing the current mechanical state, the angles of joints and tensions in the muscles, etc. Prompting yields the control signals that enable going from the present state to the next state, and the process continues. When listening to one's auditory signals, the memory is prompted to provide control signals to the vocal tract to enable the next "step" of speech. Motor control signals are not generated in real-time but are drawn from memory.

14.1 Walking

Control of muscles in the legs and in the body is a complicated process. The control signals originate in the brain and some further processing takes place in the muscles themselves. Imagine trying to compute thousands of control signals simultaneously, all synchronized to enable walking.

It is believed that these control signals are not computed in real time, but instead, they are drawn from memory. Control planning is also drawn from memory.

While walking, sensory signals from all over the body give the brain information about the mechanical state of the body. State variables are, among others, angles of

B. Widrow and E. P. Katz, *Cognitive Memory*, Springer Series on Bio- and Neurosystems 17, https://doi.org/10.1007/978-3-031-80939-2_14

the joints and tensions in the muscles. Also, vestibular signals are available, and information processed from real-time visual images is available. All this gives the brain information about balance and orientation with respect to the earth's gravitational field.

These sensory signals are real-time variables. They serve as prompts to the memory containing the recorded control signals. The control signals are not generated in real time but are retrieved from a cognitive memory. This works like a look-up table.

An interesting question is, how did the control signals get into the memory originally? They had to be generated somehow and then stored in the memory. One can only speculate about how this happens.

A newborn baby has an initial control system for walking stored in its memory. The baby begins to use this innate system after about a year when the baby becomes a toddler. The original system is inherited over generations and is generic. The system needs to be tailored to the actual physical characteristics of the baby and must evolve as the baby grows. There is a learning process that can enable this adaptivity. It is called reinforcement learning for neural networks and uses reward and punishment. When the person walking is happy with a specific movement, the contents of the memory system are rewarded. When a movement is unsatisfactory and needs improvement, the contents of the memory are punished. Learning by reward and punishment is a slow process. There is a very gradual improvement of a baby's walking which takes place during the baby's growth. Initial walking is wobbly, the walking of a toddler. Improvements are slow, but this works.

If an adult were to experience a significant leg injury, walking could be like that of a toddler. In spite of the deformity, walking could return to normal, over time. This would be like a toddler learning to walk properly.

14.2 Reinforcement Learning with a Cognitive Memory

Reinforcement learning is not unsupervised learning, nor is it supervised learning. The process is something in between the two. An example is Darwinian "learning" (i.e., natural selection). A weed growing in a lawn, if untreated, would slowly overpower the grass and take over. The fitness of weeds is stronger than that of grass. The weeds are rewarded with the ability to propagate everywhere in the local area. At the same time, the grass is punished as it is overwhelmed and overtaken by weeds.

Reinforcement learning integrates with cognitive memory to enable constantly improving walking. This works in the following way: Think of an initial walking system as a walking algorithm being stored in memory, such as with a newborn baby. The initial system is the *de facto* walking system the baby toddles with. Now let a small random change occur to the de facto system and determine if the gait improves or not. If the person senses an improvement, the improved system is rewarded and becomes the new de facto walking system. If the change results in a poorer gait, this change is punished by being deleted and not becoming the de facto system. Now another random change is tried, and so forth.

A change is only accepted if it results in perceived improvement. This process is accordingly stable. Instability would never be an improvement. The algorithm converges on a de facto system that delivers a fine walking gait.

14.3 Early Works on Reinforcement Learning

The earliest work known to these authors begins with a book by Thorp entitled "*Beat the Dealer: A Winning Strategy for the Game of Twenty-One*" [1] (sometimes called "Black jack"). Thorp developed a mathematical theory for the game. An article in Time magazine called attention to this book. It was clear from his solution that it could be implemented by using a single ADALINE neuron, if the weights were set appropriately. This was achieved by means of reinforcement learning.

The dealer performs a pre-set process shuffling a card deck. The player has several choices in the game. The final choice will be "stick." The two most often used choices are

- I will accept another card, or "hit."
- I will not accept another card, or "stick."

The player's objective is to continue to accept cards from the dealer so that the sum of the values of the accepted cards comes as close to 21 as possible but does not exceed 21. Otherwise, the player's hand results in a "bust" (loss). The player draws a card. Then the dealer draws a card. If the dealer's sum is closer to 21 than the player's sum, the dealer wins. If the player's sum is closer to 21 without exceeding 21, then the player wins. The level of performance obtained with Thorp's simplest strategy is 49.5% for the player.

How could ADALINE learn Thorp's strategy, which Thorp proved was optimal? The answer was reinforcement learning. ADALINE knew nothing of the game of Twenty-one. It was only told at the end of a game if it had won or lost. When it won, the decisions in the game are rewarded. If ADALINE loses, then the decisions of the game were punished. The input patterns to ADALINE were the binary representations of the game state, the dealer's card values sum and the player's card values sum. The decision of the player was based on the input pattern, representing the player's sum and the dealer's sum.

In the winning game, ADALINE's decisions were reinforced by adapting with the LMS algorithm. ADALINE begins play with random initial weights and plays very badly, mostly losing but sometimes winning by luck. Each decision of the winning game was reviewed, and ADALINE was adapted with the assumption that each decision, $+1 =$ hit, $-1 =$ stick, was the desired response. Each decision of the winning game was rewarded by adapting ADALINE as if these decisions were correct, giving the needed desired responses to the LMS algorithm. Each decision of the losing game was punished, by adapting ADALINE with desired responses that were opposite to the losing decisions. By rewarding the winning game decisions and punishing the losing game decisions, ADALINE slowly was trained to converge on Thorp's strategy, all while having no idea what the game was about. Training was slow because sometimes games could be won with a poor strategy, and some games could not even be won when playing with the optimal strategy. Nevertheless, reinforcement learning works.

Darwinian learning works even if the learning system knows nothing of biology; only knows if it wins or loses. Natural selection works. It is a good reinforcement learning algorithm.

ADALINE learning to play black jack (also known as the game of Twenty-one) is described in a paper by Widrow et al. [2]. At that time, the learning process was called learning with a critic. The critic tells the system whether it won or lost but knows nothing of the individual decisions. Years later, this learning process became what is now called "reinforcement learning". The original paper can be found on Widrow's Stanford website [3], clicking on publications, and clicking on the specific reference.

Another paper related to reinforcement learning is one by Widrow and McCool [4]. Random search is Darwinian [5]. Steepest descent is the LMS algorithm. The paper shows that both methods derive the same solution, only that Darwinian random search is very much slower especially when there are a large number of weights or parameters. This paper can also be found on Widrow's web site [3]. Mathematical theory and simulations are given in that paper.

14.4 Motivation and Sense of Direction

In order to walk, there needs to be a purpose and a desire to do this activity. A sense of direction is needed to steer the walking toward the objective. These are important inputs to the walking algorithm.

14.5 Speaking

Muscle control of the vocal tract to produce desired speech is a very complicated process. There are a very large number of variables to be controlled. Many of the variables have not been explicitly identified.

Imagine trying to control your tongue and your lips at the same time. Words and the sound of the words developed much later than the development of the vocal tract. (Note that the vocal folds and glottis are designed by nature to prevent food from getting into the lungs). The control algorithm for speech has to be learned and is therefore not intrinsic but has to be learned by each individual. Many muscles are involved in the control algorithm implementation. Each muscle contributes to the production of the spoken word.

When training the muscle controls, desired responses are not available for each muscle or muscle group. All that is available is the sound of the spoken word which is an aggregate effect of the activity of all the muscles. This is reminiscent of the black-jack game, where the optimal individual decisions are not known (except to the Thorp strategy). All that is known is that there was a win or a loss. As with the game of blackjack, learning to perform is accomplished by reinforcement learning.

When learning to speak, winning or losing is indicated by actions and facial expressions of the intended person. Furthermore, the listener may give insight by providing an example of the correct sound of the word that was just spoken. These samples are stored in memory and the learner could, acting alone, compare produced sounds to recorded sounds. Reinforcement learning and memory both come into play.

A baby speaks "baby talk." Listening to words spoken by others gives a large number of samples of each word. Over years of practice and with the growth and development of the vocal tract, the young person begins to speak like an adult, including local accents. Cognitive memory plays a vital role in learning to speak.

Here is a Professor Widrow story:

> I grew up in Norwich, Connecticut and apparently spoke with a "New England" accent. Some years ago, I attended a conference in Naples, Italy and made an oral presentation. After the session, a group of American attendees formed to which I joined. They were joking about my New England accent. One of the group said "You can get the boy out of Norwich but you can never get Norwich out of the boy."

14.6 Walking and Speaking

- While walking, sensory signals from all over the body deliver to the brain information about the mechanical state of the body.
  - These sensory signals act as prompts to the memory, which in turn provide muscle control signals that enable walking.
  - This works like a lookup table.
  - Muscle control signals are not computed in real time but are pulled from memory. Control planning is also pulled from memory.
- While speaking, muscle control of the vocal tract is pulled from memory in response to prompts.
  - The brain does not compute these control signals in real time.

14.7 Training Auto-Associative Networks by Reinforcement Learning

The auto-associative neural networks of Chaps. 7 and 8 were trained with the backpropagation algorithm. This was based on stochastic gradients with the method of steepest descent. The backpropagation algorithm was fast and efficient. It involves the transmission of signals through the weights in the forward direction to derive network outputs and transmission of error signals through the same weights in the reverse direction. i.e.: backpropagation, in order to achieve error signals for adaptation and learning. This algorithm is the basis of the current AI revolution.

The problem with the backpropagation algorithm is that it seems not to be biological. It is hard to imagine nature developing a method for sending signals both ways through the same synapses.

Another method for adapting an auto-associative neural network that would be more biological-like is based on reinforcement learning. The objective is to reduce error with each adaptation cycle in a natural way. Here is the idea:

> Begin with random initial weights. The initial mean-square-error would probably be huge. Now make a random change in the weights. Check the mean-square-error and if it is smaller, reward the weight change by adding it to the weights of the network. If when making the random change and the mean-square-error has increased, punish the random change by subtracting it from the network's weights. This is a slowly converging process, much slower than backpropagation, but it is simpler to implement, something nature could do.

Another method that would be even simpler for nature, but slower yet, would be to accept the change to the weights when rewarding, and to delete the change and go back to the weights before change when punishing. Both methods are very slow compared with backpropagation, but they are much simpler. Both methods are stable because they make the mean-square-error smaller with each adaptation cycle.

These are random search algorithms, related to the algorithms studied by Widrow and McCool [4]. Even though they are slow, they could be used for training the neural networks in the memory segments.

Storage of new data is fast, keeping up with incoming data streams. Slow neural network adaptation means that retrieval of new data from long-term memory may not be possible initially. This can be done only after convergence. (Temporary retrieval from short-term memory is possible.) Once the memory of a segment is full, its neural network will slowly adapt and stabilize, not needing change thereafter except to train itself in the rare event of component failure in the network. The neural network can adapt around its own internal weight failures. If access to newly recorded data is needed, that could be obtained by accessing the short-term memory with prompting.

Here is another B. Widrow story:

> Some years ago, I met my friend Paul Werbos (the inventor of the backpropagation algorithm) at a conference, and we had some good chuckles. I told him that every year new learning algorithms are published that show performance superior to LMS. Year after year, every

such paper compares the new performance to LMS. The benchmark had become LMS. Paul told me that every year new learning algorithms for neural networks are published that show performance better than backpropagation. The benchmark had become backpropagation. With all the new "better" algorithms proposed each year, the algorithms that are universally used today are LMS and backpropagation.

On another subject, Paul and I discussed the possibility that backpropagation would be nature's learning for neural networks. We both agreed that this is unlikely. That backpropagation is an excellent algorithm for engineering applications, but not for nature.

14.8 Summary

Control signals for walking and speaking are not generated in real-time but are drawn from memory. How did the control data get into memory, how was it generated? Babies are born with intrinsic walking control systems that are generic, not tailored to the individual baby. The inborn system is good enough to start but over time is improved by reinforcement learning. The system adapts to the baby and continues to adapt as the baby grows. Control of the vocal tract does not begin with an initial control system. The vocal folds and glottis were not designed by nature for speaking; they were designed to allow breathing and to prevent food from entering the lungs. Speaking begins with random sounds and improves slowly with reinforcement learning.

Appendix

Use google (or your favorite internet search engine) to explore:

- Motor cortex memory
- Prompting the motor cortex for walking and speaking
- Motor control and learning
- Walking, speaking control signals
- Reinforcement learning
- Wikipedia, first paper on reinforcement learnings
- Wikipedia, search for Darwin
- Wikipedia, search for reinforcement learning.

References

1. Thorp, E.: Beat the dealer: a winning strategy for the game of twenty-one: a scientific analysis of the world-wide game known variously as blackjack, twenty-one, vingt-et-un, pontoon, or van-john. Random House, New York, ISBN: 9780394703107, 0394703103. OCLC: 655875, 220 pages (1966)
2. Widrow, B., Gupta, N.K., Maitra, S.: Punish/reward: learning with a critic in adaptive threshold systems. IEEE Trans. Syst. Man Cybern. **3**(5), 455–465 (1973). https://doi.org/10.1109/TSMC.1973.4309272
3. Widrow, B.: Publications. https://isl.stanford.edu/~widrow/publications.html
4. Widrow, B., McCool, J.: A comparison of adaptive algorithms based on the methods of steepest descent and random search. IEEE Trans. Antennas Propag. **24**(5), 615–637 (1976). https://doi.org/10.1109/TAP.1976.1141414
5. Darwin, C., Kebler, L.: On the origin of species by means of natural selection, or, the preservation of favoured races in the struggle for life. J. Murray, London (1859). Retrieved from the Library of Congress, https://www.loc.gov/item/06017473/.

Chapter 15
Pattern Recognition with Features and Whole Images

Abstract Hubel and Wiesel's work with the visual systems of cats presents positive evidence that pattern features are extracted in the visual cortex and are no doubt critical for pattern recognition. Researchers working on pattern recognition systems were strongly influenced by this work. Features are important. On the other hand, research of Julesz using random dot stereograms shows that visual processing can be done with featureless random dot patterns. These ideas of pattern recognition seem to conflict. A singular conclusion is that the animal visual system uses both features and the whole pattern to process visual images. An approach to pattern recognition design would benefit from analyzing both features and whole images.

There are two approaches in the field of pattern recognition. One method is based on feature detection and analysis. The other method analyzes the entire pattern, without looking for or using features. Examples of features would be straight lines, circles and other geometrical elements present in an image. These could be detected and used as inputs to a recognition system.

The face recognition system described in Chap. 11 resembles practices using both methods. It is a two-mode process. The first mode trains a generic face detector and uses it to locate a human face in a complex scene (this face detector could be considered to be a feature detector). The second mode uses the location of a detected face to begin analyzing the facial image. The analysis is of the entire facial image. It is not based on biometrics that would, for example, measure the distance between the eyes, the distance from an eye to the tip of the nose, etc. The biometric measurements are features that could be inputted to a recognition system. A use of such features is a good idea but is not incorporated into the method of Chap. 11.

B. Widrow and E. P. Katz, *Cognitive Memory*, Springer Series on Bio- and Neurosystems 17, https://doi.org/10.1007/978-3-031-80939-2_15

15.1 Hubel and Wiesel's Cat

In 1981, David Hubel and Torsten Wiesel received the Nobel Prize for physiology or medicine for their discoveries of the visual system and visual processing. They recorded electrical activity from individual neurons in the brains of cats. They found that certain neurons in the visual cortex only fire when a horizontal line is being seen. Whereas other cells of the cortex only fire when seeing a vertical line. Other cells only fire in response to a line at an oblique angle. They examined the effects of line, squares or rectangular stimuli. They identified neurons that specialize by responding to these unique stimuli.

The discoveries of Hubel and Wiesel have had a profound effect on the field of pattern recognition. Researchers became very interested in feature detection and would focus their research in that particular direction. Excellent results were obtained with this approach. Nevertheless, not all work on pattern recognition uses features. The entire pattern could be thought of as a single feature.

15.2 Julesz and Random Dot Patterns

Béla Julesz, a Hungarian American, was well known for his invention of random-dot stereograms [1]. As a MacArthur fellow, he directed the laboratory of Vision Research at Rutgers University. His primary research area involved Neuroscience and psychology specializing in the study of binocular vision and depth perception.

During the Soviet occupation of Hungary, Dr. Julesz left for the United States. He worked at Bell Laboratories on binocular vision where he developed the random-dot stereogram to facilitate his study into depth perception. This effort culminated in the publication of his 1971 book, "Foundations of Cyclopean Perception" [1]. Retiring from Bell Laboratories in 1989, he joined the Rutgers University Faculty.

While at Bell Laboratories, he had a sabbatical year that gave him the opportunity to travel across the U.S. visiting various universities. During this trip, he came to Stanford University to deliver a seminar on his work. Widrow attended. He passed out paper eyeglasses with polarized lenses. He was able to project two different polarized random-dot images simultaneously. A vertically polarized image addressed to the left eye and a horizontally polarized image addressed to the right eye. Three-dimensional objects composed of random-dots were perceived. Each eye only saw random dots, as could be confirmed by looking with one eye closed. The three-dimensional effects were spectacular. Many different subjects were shown. Widrow remembers seeing a rectangular hole in the ground with a ladder in it, obviously to enable one to climb down to reach the bottom. Widrow remembers seeing a corkscrew popping out from the screen that made him duck to avoid getting hit by it. The lecture was well received. After the lecture, there was an opportunity to ask him one-on-one questions. During his lecture, he described experiments that he conducted during a stay at Harvard University. A Harvard student with a photographic memory (attending Radcliffe college) visited him to try out some ideas. She could look at a random dot pattern for the left eye with her right eye closed. A week later, she looked at the random pattern for the right eye with the left eye closed. Julesz asked her what she saw, she described seeing embossed letters. He asked her what do they say? She answered, "Now is the time for all good men to come to the aid of their country."

At the outset, all she saw before her was the right-eye random dot pattern. She was able to visualize the three-dimensional effect by remembering the random-dot pattern seen the week before. Combining these images, she could then see the words embossed in three dimensions. She was able to do pattern recognition with featureless patterns, and would need to be able to recall featureless random-dot patterns.

Widrow asked if the holding of the patterns in position was critical. Julesz said no, she could get the effect with the images at different distances and not needing to orient the images at a precise angle. The patterns that she saw were featureless random dots, and she would recognize objects with them. Widrow asked Dr. Julesz, "How many people have this kind of photographic memory?" He suggested a guess of perhaps 10–100 out of a billion. Widrow then asked, "Is she brilliant?" and the reply was "yes". Widrow finally asked, "Is she happy?" The reply was that she was very unhappy.

What is the "take away" from this presentation and demonstration of an extraordinary feat of memory? It described interesting properties of human pattern recognition, such as within reason insensitivity to the size (scale), translation, rotation, aspect angle, etc. Insensitivity to moderate aberrations. The memory feat, remembering in exquisite detail entire images that are random dots, and featureless. Being able to process them in a binocular way to derive depth perception was amazing.

Some years later, Bernard Widrow bought Julesz's book and studied Julesz's work in greater depth. Teaching his classes about this work, he needed to explain how three-dimensional perception works. He created an idea of a polar bear on a snow-covered ice floe.

15.3 The Polar Bear

Suppose it were possible to hover about five meters (about 16.4 feet) above a polar bear located in a featureless snowy landscape. The bear is white like the snow. The bear and snow would blend rendering the bear essentially invisible. With a bucket of black paint and a small brush that drips, coat the entire scene, the background snow and the bear, with black dots. Now the dots on the bear would be visible sticking up from the background and being on a large three-dimensional object having the shape of a polar bear. The bear would still be invisible, only the black dots on the bear would be visible.

At five meters above, the background's dots would be visible to both eyes. The position of the dots on the bear relative to the background dots would be different for the two eyes. Imagine the geometry and the projections.

A single dot suspended one meter above the background would be one meter closer to the observer than the background dots. The two eyes would see different images of the dot above the background dots. If the dot were suspended two meters above the background, it would be two meters closer to the observer.

The closer the observer is to the bear, the more pronounced will be the three-dimensional effect, and the more dissimilar will be the images observed by the two eyes. The ratio between the bear's height above the background plane to the distance of the observer determines how pronounced will be the three-dimensional effect. At a distance of, for example, 50 m, the images of the observer's eyes would be essentially the same. The bear and its dots would blend into the background preventing the bear from being seen.

For the binocular effect to provide three-dimensional perception, the images of the background dots would need to be seen as identical, even though the position and scale might differ from eye to eye. This correlative processing is performed on patterns that have no features. The processing is done on the entire background images. There is no feature detection, but there is three-dimension recognition.

By programming his computer to generate random dots for the left eye and random dots in the same position for the background and some dots displaced relative to the background, Dr. Julesz was able to create random-dot stereograms. These would be the patterns seen by the two eyes looking down on the polar bear.

Some years later, Prof. Widrow gave a lecture on cognitive memory to a group of people at U.C. Berkeley. The audience consisted of engineers, neuroscientists, and psychologists. When the lecture was finished, one man in the back row raised his hand and said that he knew the young woman who was Dr. Julesz's research subject. She could read a book and repeat perfectly the words on any page when she was given a page number. She graduated from Radcliff, worked for a few years, married, and had a baby. As soon as the baby was born, she noticed a change. She no longer had a photographic memory. She was much happier.

15.4 Jill Price

In 2008, on the television program Good Morning America™ (ABC Broadcasting Company), Diane Sawyer interviewed a 42-year-old woman who had a photographic memory. Her name is Jill Price, and she was a research subject with a group of neuroscientists at U.C. Irvine. At age 14, she began to keep a diary that, at the time of her interview, had 50,000 pages. She recorded events in her life and news events. The pages were numbered. Given a page number, she would recite what she had for lunch that day, and who she saw throughout the day.

Give her a date and ask her what happened that day and she knew. Given the date of the fall of the Berlin wall, she could say what she had for lunch, who she saw that day, but nothing about the Berlin wall. Given another date and asked what happened? She said on that day Elvis Presley died.

When she was a schoolgirl, she could not remember poems and other schoolwork. She was only able to remember what she wrote in the 50,000 pages of her diary. She was very unhappy with her photographic memory.

15.5 Vaughn Smith

On July 2, 2022, the Washington Post reported a story about a 46-year-old man named Vaughn Smith, an autistic savant. Mr. Smith could converse in 24 languages. He is familiar with a total of about 60 languages. A person who speaks many languages is called a polyglot. A person who speaks more than 10 languages is called a hyperpolyglot. That describes Mr. Smith.

When he was a child and then a young man, he seemed to be a misfit. He had trouble relating to people. But not all people. He learned languages by interacting with people from all over the world. For example, he befriended a group of Dutch travelers in a Starbucks, and they couldn't believe that he had never been to Holland. He lived in Washington, D.C., and while attending schools with international students, he learned to speak these languages with the correct local accents.

As a schoolboy, he was able to remember names, dates, and sounds better than most people. His extraordinary memory enabled him to learn languages. He learned American sign language from students at Gallaudet University.

He visited the Brain and Cognitive sciences lab at MIT to have an FMRI scan of his brain. While reading in English, the FMRI images were recorded. His language processing areas of the brain were smaller than normal, a real surprise.

Mr. Smith had a photographic memory that seems to be tuned to language discovery. Words representing objects and concepts are arbitrary, they have no features. Recognition of sounds of words seems based on entire words, not identifiable features.

15.6 Photographic Memory

People normally do not have photographic memories. However, the above examples confirm that the brain with its sensory inputs has the capability of photographic memory, we all have this capability but do not exercise it. A photographic memory allows one to remember visual and auditory images in exquisite detail. These images can be processed as well as current images.

Photographic memory is not a good thing. With it, nothing can be forgotten. Remembering everything requires an enormous memory capacity. We all have this capacity, but we only use a small fraction. A photographic memory is not normal. Something is broken to cause it. But what?

It is hypothesized that something is wrong with the *interesting?* switch, instead of blocking the recording of new images and data unless they are "interesting," this switch passes everything. Although the recorded database is enormous, parallel processing of a cognitive memory results in access time that is normal. Storage time is also unaffected since recording is continuous.

It is hard to imagine what living with a photographic memory would be like. Most people who have one seem to be very bright but unhappy.

15.7 Features or Not

Hubel and Wiesel's work suggests that feature detection and usage is an integral part of visual processing and recognition. There is no doubt that this is correct. On the other hand, Julesz's work, of which there also is no doubt, suggests that the entire visual pattern is processed by the visual system without features. The only conclusion that one could draw from this is that both methods are incorporated into visual processing at the same time.

15.8 Professor Widrow's David Hubel Story

> In 1987, DARPA initiated a study of neural networks throughout the U.S and also in a few other countries. DARPA contracted with MIT's Lincoln Lab to do the study. Lincoln Lab enlisted Professor Widrow to be the study director. Lincoln Lab staffers did dozens of interviews of neural network researchers. Every other month, there would be a meeting with the staffers to summarize what they have found and from this created a set of overhead transparencies.
>
> The following day, a presentation would be given from these transparencies to a very distinguished panel. Among the panel members were two former directors of DARPA, two former directors of Lincoln Lab, Lincoln Lab staffers, and others. Also on the panel was David Hubel. To introduce the subject during the first presentation, I started describing ADALINE as an example of an artificial neuron, and the LMS algorithm as a learning method. (The name ADALINE is an acronym for ADAptive LInear NEuron. See Chap. 4 of this book.)
>
> When the presentation was over, a pleasant conversation with David Hubel took place. He was a neuroscientist and professor at the Harvard Medical school, was easy to talk to, and was a most interesting person. He related his observation that ADALINE is much more than just a neuron. ADALINE is a neuron together with integrated synapses that are part of it. This was a surprise as ADALINE up to that point was only considered an isolated, artificial neuron. (He was obviously correct.)

15.9 Summary

The discovery of neurons in the cat visual system that are responsive to vertical lines, horizontal lines, and other features presents convincing evidence that features are highly important for visual processing and pattern recognition. Other convincing evidence by the work of Béla Julesz with random dot patterns shows that visual processing is performed on the entire image. Julesz's random dot patterns are featureless. The animal visual system evidently uses both features and the entire image for visual processing and pattern recognition. The combination is very effective. Modern engineers emulate this in the design of their pattern recognition systems.

Appendix

Use google (or your favorite internet search engine) to explore:

- Hubel and Weisel cat experiment
- Hubel and Weisel feature detection
- David Hubel, "Brain and Visual perception," Google books, (Oct. 2004)
- Feature detection for pattern recognition
- Biometrics
- Biometrics for face recognition
- Béla Julesz
- Random dot stereograms
- Three-dimensional imaging with random dot patterns
- Binocular vision.

Reference

1. Julesz, B.: Foundations of Cyclopean Perception. MIT Press (1971)

Chapter 16
Memory Pattern Storage and Retrieval

Abstract Images, sounds, patterns, data are stored permanently in long-term memory, cognitive memory. Parts of the brain involved with memory consist of neural networks. The general belief in neuroscience, accordingly, is that data, patterns, etc. are stored in neural networks of the brain in the form of "memory traces." This is logical but very likely incorrect. The wiring of the neural networks is constantly changing, with neurons added in the young and neurons disappearing in the elderly. These changes would cause the networks' output signals to become meaningless, unless the networks are constantly trained. Training patterns would be needed. Why would nature store patterns and data to keep training the networks in order to store patterns and data? "Catch 22." The role played by the memory's neural networks is retrieval, not storage.

It seems likely that when patterns, images, sounds, etc., are stored in memory, they are not stored in the brain's neurons and synapses. This is contrary to what most neuroscientists currently believe. Many neuroscientists accept the idea that memory is stored in the neurons and synapses in the form of "memory traces." Simple arguments cast doubt on this memory trace hypothesis.

16.1 "Memory Traces"

Beginning in 2008, a series of lectures were given by Dr. Michael Merzenich, an eminent neuroscientist, Professor Emeritus of University of California, San Francisco (UCSF). The series titled "Brain Fitness Program" was broadcast over Public Television (channel 9 in San Francisco). He is a Kavli Laureate and is recognized as "the father of neuroplasticity," and is a strong proponent of the memory trace hypothesis. His research involves learning processes, neurological illness, and the cerebral cortex neurological processes. Understanding how the brain can rewire itself, he was able to develop remedial methods for people with speech, language, and reading deficits. Dr. Merzenich first began his studies taking neurological measurements with

B. Widrow and E. P. Katz, *Cognitive Memory*, Springer Series on Bio- and Neurosystems 17, https://doi.org/10.1007/978-3-031-80939-2_16

micro-electrode mapping in animal brains. He is an inventor who has aided in the development of medical devices such as the cochlear implant.

From his Brain Fitness lectures, the following are some take-aways:

- Memory is stored in the wiring of the brain.
- Hebbian learning—Neurons that "fire together, wire together."
- Change strengthens connections between neurons.
- Memory is crucial for learning.
- Motivation is a key factor in brain plasticity.
- In case of brain injury, the brain can rewire itself over time to compensate for the loss.
- With loss of a sensory input (e.g., going blind), this results in the visual cortex being taken over by other senses such as hearing, touch, etc., making their responses heightened and more acute.

Dr. Merzenich is not only a theoretician, but he is a practical man with much experience with brain measurements and inventive technology. He has contributed strong hypotheses about brain plasticity. There is evidence that disuse of a cortex will cause its function to be assumed by other senses. There is evidence that an injured brain can, within limits, heal itself. *But where is the evidence that memory is stored in the neurons and their synaptic connections?*

16.2 Memory Stored in the Brain's Neural Networks?

There may not be direct evidence for this, but logic and reasoning seems to enhance the belief in memory traces. Logic and reasoning suggests that when neurons and synapses are destroyed, as with Alzheimer's disease, memory is lost. This implies that memory is stored in the neurons and synapses which is the widely accepted view in neuroscience and supports the memory trace hypothesis.

16.3 Memory is not Stored in the Brain's Neural Networks

We disagree with the idea that memory is stored in the neural networks. Our reasoning is the following:

- Living neural networks are subject to change.
- Neurons, synapses, and dendrites die out daily in the elderly. Neurons are added daily in the very young.
- Neural wiring changes constantly. In order to perform consistently, the wiring must be trained regularly. Without constant training, neural network functioning will become inconsistent and over time would become useless. Training can take place while awake and during sleep.

- Living neurons and synapses are analog devices and are not suitable for long-term memory storage.
- Old knowledge and new knowledge must be retrained in order to be maintained. Training in new knowledge disturbs the learned responses to old knowledge. The only possibility is training both old and new together. It is necessary to store the training patterns in order to do this.
- Would nature store the training patterns to train the neural networks to store the training patterns? Catch 22!
- Long-term memory is digital information and solid. Wake up from a deep sleep, anesthesia, or coma, and memory is intact and still works.
- Long-term memory is not stored in the neural networks. The neural networks are used for memory retrieval, not storage.

16.4 Memory with Neural-Network Defects

The ideas expressed here are based upon experience with artificial neural network hardware. It has been found that if portions of the circuits fail, the remaining circuits automatically take over the functions of the defective circuits, as long as the networks are continuously trained. Thus, artificial neural networks can adapt around their internal flaws. Network capacity is reduced with neuronal failure. Network plasticity allows network functioning but at a reduced level.

Alzheimer's patients could be asymptomatic for ten or twenty years as the disease progresses. As biological neural circuits die out, the intact circuits assume their function up to a certain point, as long as training takes place continuously. On a bad day, the patient can remember very little. On a good day, the patient remembers many things that could not be recalled the day before. Long-term memory is digital and permanent. If this was not the case, the patient would not be able to remember things even on a good day. This strongly suggests that Alzheimer's patients cannot remember things because of the ultimate failure of the neural networks involved with memory. The memory is digital and stays intact. But retrieval becomes more and more difficult because so much of the neural network is diminished that retrieval becomes problematic.

Thus, what we learn from Alzheimer's patients is that the brain's neural networks are involved with memory, but the involvement turns out to be not so obvious.

One further comment about the brain "rewiring" itself when there is some form of impairment. It does not really rewire. The wiring stays the same. Only a limited number of defective neurons and synapses drop out of the circuit. The remaining, intact components adapt by changing synaptic weights, thus compensating for the loss. This is one form of brain plasticity.

16.5 Summary

Long-term human memory is stored digitally, not in analog neural networks. The purpose of the memory's neural networks is retrieval, not storage. Alzheimer patients lose neurons and synapses, and thus, experience memory loss. Data and images are stored permanently in long-term memory and remain intact, even with progression of Alzheimer's disease. The reduced capacity of the neural networks causes retrieval to be difficult or impossible, resulting in "memory loss."

Appendix

Use google (or your favorite internet search engine) to explore:

- Long-term memory storage [Wikipedia]
- Long-term memory retrieval
- Adding neurons when young
- Losing neurons when old
- Losing neurons due to Alzheimer's disease
- Michael Merzenich.

Chapter 17
Pattern Retrieval and Dreams

Abstract Dreaming takes place during REM (rapid-eye-movement) sleep. All sensory inputs are turned off. The source material or subject matter of a dream (images, data, etc.) can be obtained from memory. As the dream is being composed, it is stored in both long-term and short-term memory, just like streaming sensory inputs would be stored. If awakened in the middle of a dream, the dream thus far will be recalled from short-term memory in full detail. If awakened post-REM, the dream will not be recalled. There will be no awareness that dreaming had taken place. Recall of the dream from long-term memory would not happen since the necessary prompt would be unknown. Searching for the prompt is done by psychiatrists talking with a patient attempting to retrieve "bad dreams" from long-term memory, all for the purpose of therapy.

Long-term memory patterns are stored in the folders of the segments. Each segment has an autoassociative neural network that is trained autonomously. When prompted, it indicates *déjà vu*, or not.

17.1 Training the Neural Networks for Data Retrieval

The neural networks are trained with the patterns or images stored in the folders. If the answer to the *déjà vu* is yes, there is a hit and the contents of the hit folder are discharged onto the memory output line. A hit means that the prompt pattern or image bears a relation to at least one pattern in the respective hit folder.

Although the images from the various sensors are recorded in the same segment when "interesting," they have separate folders and separate neural networks. If any of the sensory prompts causes a hit, the entire contents of the folder for each of the sensory signals will be discharged as memory output. For instance, the smell of cooking could cause a person to recall the image of their mother preparing food in the kitchen.

B. Widrow and E. P. Katz, *Cognitive Memory*, Springer Series on Bio- and Neurosystems 17, https://doi.org/10.1007/978-3-031-80939-2_17

During a two-person conversation, the speaker is interrupted as something that was said has prompted the other person to interrupt. The interruption is urgent because the prompt, stored in short-term memory, will likely soon be forgotten by the other person. Interruptions are frequent during most conversations sometimes causing the subject to drift from topic to topic.

Training of the autoassociative networks is autonomous, i.e., unsupervised. All that is needed are the patterns stored in the folders.

Training can take place when awake and when sleeping, particularly during REM (rapid eye movements) sleep. Sleep has stages 1, 2, 3, etc., and periods of REM. Dreaming takes place during REM [1].

17.2 Sleep and Dreams

A typical sleep cycle lasts for about 90 min. It begins with the various stages, non-REM, and then goes into REM. During non-REM sleep, all of the sensory inputs are disabled. The brain transitions into quiet mode, and the training of the neural networks is probably slowed. During REM sleep, the brain is active, similar to a wakeful state. Although all sensory inputs are disabled, input patterns and images are supplied from memory instead of from the body's senses. Training during REM probably takes place in a very similar manner to when awake.

The images sensed during a dream are very vivid and could even be somewhat surrealistic. Different thoughts are combined in unusual ways, and this leads to creativity.

Eye movements are detected with electrodes attached above and below the eye. The eye muscles that move the eye generate signals that are easily detected. The eyes are moving because, with the reality of the dream, the dreamer is conversing with someone in the dream. Not a sound is made, however. During REM, the entire body is paralyzed except for heartbeat, for breathing, and curiously for eye movements.

Creative solutions generated while dreaming are most often lost. They are stored in both long-term and short-term memory. To be stored in long-term memory, the dream must pass the "interesting" test. Everything is stored in short-term memory without testing, albeit temporarily and does not persist.

Dreams can be retrieved from long-term memory if they were recorded, and if the prompt were known. Dreams can be retrieved from short-term memory if the dreamer were awakened in the middle of dreaming. The dream itself serves as the prompt. Recollection of a dream is generally a matter of retrieval from short-term memory. If the dream is sensed to be important, preserving it would require immediate recording in long-term memory. Otherwise, it will fade from short-term memory.

When awakened in the middle of the dream however, the dreamer can describe the dream thus far in detail.

After a 90-min sleep cycle, awakening when not dreaming will render the sleeper unable to recall anything about dreaming. Dreaming had taken place, but the sleeper is unaware of this because the sleeper would not know the prompt.

A Professor Widrow story:

> Some years ago, I woke during the middle of the night and realized that in my dream I had uncovered the solution to a control theory problem which had eluded me for more than 10 years. I immediately reached for a pen, a yellow pad, and quickly wrote down the solution before forgetting. The next morning, I called my postdoctoral scholar, Eugene Walach, described the solution, and asked him to simulate and test this new idea. He did, though he had to make a small change moving a signal from a pre-summation point to a post-summation point, and then it worked. The source of the solution had come from the dream. The result is that with great enthusiasm we wrote a control theory book together, "Adaptive Inverse Control" [2].

The subject of talk therapy seems related to dreaming. The patient reclines on a sofa and talks with the therapist. The patient had a terrible experience sometime before and does not remember all of what this is although stored in long-term memory. The hope is that by talking with the therapist the discussion will by chance hit upon the prompt. Not knowing what the prompt is could take years to discover. If the particular prompt is finally found, the patient and the therapist could discuss the experience and thus try to neutralize it.

Post-traumatic stress disorder (PTSD) patients are among the patients receiving this kind of therapy. Psychedelic drugs are now used to put the patient into a kind of dream state in order to hasten the hoped for discovery of the unknown prompt [3].

Another Professor Widrow story:

> Some years ago, my wife and I were in San Francisco and visited an exhibition at Fort Mason of beautiful Mexican jewelry that was for sale. My wife and I were viewing one of the showcases when I spotted a young woman across the room at another showcase who was exquisitely beautiful. It was strange. One of her arms was in a sling. How could anyone so perfect have something broken? I was curious, so I slowly approached from behind her and asked her about the arm. She said that she is a sleepwalker and suffered the injury by tripping and falling during a particular nocturnal occurrence. I asked her how she knows that she is a sleepwalker. She said that she could go to the refrigerator in the middle of the night, take out something to eat, and then go back to bed. The next morning she would find some perishable food on the kitchen table. She knew that she would never leave that food outside the refrigerator if she were completely awake. She had no memory of walking in her sleep. She was in non-REM sleep. In REM sleep, she would not have been able to move due to temporary sleep induced paralysis. But her sensory inputs were not blocked and must have functioned properly since she was able to find the refrigerator and the food kept within. This is strange. She was able to be in non-REM sleep, but her sensory inputs were working and she was thinking, as if being awake, but recording none of the sensory input or actions taken.

A take-away from these stories is that neural networks are training while awake, while in REM sleep, and possibly while in non-REM sleep.

It is interesting to note that EEG recordings show patterns during REM that are similar to patterns recorded while awake. EEG patterns recorded during non-REM sleep are quite dissimilar.

17.3 Awake

When we are awake, all of the human body's senses are providing signals. They are on and functioning. All new sensory input data enters the memory system. This data remains in long-term memory if, when it is examined, it is determined to be "interesting." If not, this data will be blocked and eventually fade. If deemed to be interesting, the data will be recorded in long-term memory for permanent storage.

Searching long-term memory to try to remember something is actually trying to find the correct prompt. Prompts come from the "environment," sights, sounds, etc. streaming in from the senses. Prompts also can come from data derived from an already prompted folder. Since there are many items of data stored in each folder, a prompted folder can provide a variety of additional prompts related to the original prompt. This is known as "prompt feedback."

Sometimes prompts can be noisy. The result will be a recollection of data that "pops into your head" that was not being sought. Data recorded long ago and not accessed for a very long time can suddenly and spontaneously be recalled. This is always surprising and sometimes annoying. Data that is frequently accessed leaves many prompts, making subsequent recall easy.

Recovery of data stored in short-term memory is readily accessible, but a prompt is needed. Things just recently learned can be easily recalled. Inputs that occurred beyond the memory time of short-term memory can be recalled from long-term memory only with the proper prompt. Recall from short-term memory is based on when the data was inputted. Some perception of when things happened is the key to finding short-term data. Once such data is retrieved prompting results in retrieval of specific pieces of data.

Recalling things that occurred just after the time span of short-term memory, for example a day later, can often be recalled easily. The incident recorded creates an environment or is created by an environment, so that the prompt can come from the environment once again.

Things recorded over the past month or two can be recalled readily by a prompt derived from a persistent environment or derived from having been recalled recently. Recalling creates an environment of new prompts. Recorded data is re-recorded into new folders. Recollection can be achieved, with prompting, by searching the memory folders recorded in the past few months. Like with short-term memory, search is assisted and directed by some perception of time associated with the occurrence.

Data recorded long ago and not recently recalled can only be accessed with a correct prompt. This will only come from something in the environment. The environment might also include the voice of another person making comments or asking questions.

17.4 Asleep

During non-REM sleep all sensory inputs are suppressed and the brain is essentially off-line to the external world. Nothing will be recorded in memory. Training with the data already stored in the folders could continue. One cannot be sure of this, however, and more research may clarify this issue.

During REM sleep, the sensory inputs are also suppressed. Memory activity is practiced using data stored in the folders rather than with the arrival of new data streaming. But, the brain is on-line, similar to the wakeful state.

Recalling a folder during REM often begins spontaneously without an initial prompt. Patterns from the retrieved folder could become prompts to retrieve more folders in a forward-chaining manner. The retrieved data are then combined with a thought process to comprise a dream. The dream is then stored in short-term memory and if interesting in long-term memory. The evidence is the experience of a sleeping person awakened while having REM and being able to recite vivid elements of the dream. The purpose of dreaming appears to be problem solving. Solutions that may be useful can be recovered subsequently when awake, having been woken up during REM.

Although most dreams are about random subjects, having been initiated spontaneously, some dreams are a continuation of the thought process ongoing while falling asleep. Such thought processes, if recorded in short-term memory, could contain prompts that would cause the continuation of thought process while entering sleep. Dreams recorded in short-term memory must be captured quickly when woken up, otherwise they will fade and be discarded.

It seems that some of the best thinking occurs when half-awake, half-asleep, when falling asleep and when waking up. Dreams are usually fantasies, but sometimes a good idea might come from a dream.

17.5 Schizophrenia

Schizophrenia is a disorder that affects a person's ability to think, feel, and behave clearly. This is a long-term mental disorder involving a breakdown in thought, emotion, and behavior with faulty perception, withdrawal from reality, and living in a dream world of fantasy and delusion. Patients with schizophrenia can experience hallucinations and delusions, hearing voices of people not present, having incoherent speech, and flattened emotions. Schizophrenia patients in a quiet room can have conversations with people, speaking to and hearing them, with no one else in the room.

It is believed that a schizophrenic patient, when hallucinating, is asleep and awake at the same time. They are experiencing dreams superimposed on top of reality. This is hopelessly confusing. Since they can speak and move about, they are awake. But at the same time, they are having REM sleep. The dreams are totally vivid. This is an incurable disorder, so far.

A person with schizophrenia can be awake and in REM sleep at the same time. The schizophrenic does strange things while awake. The sleepwalker does strange things while sleeping.

17.6 Summary

Both sleep walkers and schizophrenic patients can experience being asleep and awake at the same time. The sleepwalker walks with non-REM sleep, and the schizophrenic patient during hallucinations is having REM sleep. The sleepwalker can walk but not record experiences or recall them when normally awake. The schizophrenic patient records the dream conversations and could possibly recall this if disturbed in the middle of having a hallucination. Normal people do not record during non-REM sleep and with no sensory inputs, would have nothing to recall. During REM sleep, dreams are recorded in short-term memory and can be recalled if awakened during the dream. Memory retrieval is very much dependent on the recording circumstances, and finding the right prompt.

Appendix

Use google (or your favorite internet search engine) to explore:

- REM sleep
- Sleep stages
- Sleep walking
- Schizophrenia.
- Prompting recall from dreams
- PTSD
- William Dement

References

1. Kryger, M.H., Roth, T., Dement, W.C.: Principles and Practice of Sleep Medicine, 6th edn. Elsevier, Amsterdam (2015)
2. Widrow, B., Walach, E.: Adaptive Inverse Control. IEEE Press, New York, NY (2008)
3. The rebirth of psychedelic medicine. Stanford Science Digest, Discovery (1 Dec 2023)

Chapter 18
Innate Knowledge

Abstract Innate knowledge is knowledge that a person (or animal) is born with. This knowledge resided in the DNA of the single first cell. With development, this knowledge was stored in the DNA of all subsequent brain and body cells. Innate knowledge is inherited. It can be passed onto future generations. For a father, inborn knowledge in the DNA is passed down through sperm. For a mother, the inborn knowledge is passed down through an egg's DNA. Knowledge gained after birth is not passed down to the next generation. This knowledge is stored in brain cells, not in all other body cells, and the only way for this knowledge to propagate from one generation to the next is through schooling. An example of innate knowledge of mammals is the control system for walking. A baby deer walks within a few minutes of birth. Like the baby deer, the human baby has the knowledge of walking installed in the developing brain before birth. The human fetus practices walking by kicking. At birth, the human baby does not walk. Once the leg muscles are developed enough, the baby walks. The walk algorithm is based on innate knowledge stored in long-term memory.

18.1 Examples

Inborn knowledge in the form of patterns, images, data, etc. is pre-loaded in the developing brain's memory and remains intact throughout a lifetime.

Examples of inborn knowledge are the following:

- A bird building a nest involves complex construction in "safe" places such as bell towers, roof tops, treetops, telephone poles, etc.
- A baby bird knows how to fly. When physically developed with feathers and wing muscles, the baby jumps from the nest. It must fly or crash into the ground. It could not have learned to fly before crashing. The knowledge must have been in-born. The neural control signals that enable flapping of the wings are highly complex and could not have been learned by simply observing the flight of its parents.

B. Widrow and E. P. Katz, *Cognitive Memory*, Springer Series on Bio- and Neurosystems 17, https://doi.org/10.1007/978-3-031-80939-2_18

- A baby deer is walking within minutes of birth. It could not have learned to walk in a few minutes. Knowledge of how to walk must be in-born.
- A human baby born sucking, crying, peeing, and pooping. It could not have learned this in utero. It must have been in-born and stored in long-term memory.

An amazing example of in-born knowledge is that exhibited by the new-born kangaroo. The kangaroo belongs to the marsupial family characterized by the mother's pouch where the baby, called a joey (Aboriginal for "little animal"), stays during continued development after being born in a very immature state. The birth of the joey is premature; so, the little creature is almost a fetus just after birth. It has tiny forelegs and slightly larger hind legs, and is born blind. This tiny, blind fetus with its tiny legs crawls all the way up into the mother's pouch where it firmly attaches to a milk-giving teat to complete development. The joey was born with full knowledge of how to do these amazing things without help.

A Prof. Widrow story:

My friend, James Spilker, had a home overlooking the Pacific Ocean set on a large property of about 100 acres populated by a wide variety of wild animals. Jim told me this story:

> One day, he and his wife were near a large window, looking out. A deer was outside, standing still. This was most unusual since deer never stand still. They were watching when a baby deer began to emerge from the backside of the deer. The baby landed on the ground, and the mother turned and licked the baby. Then the mother began to walk away and the baby got up and followed her. The baby deer could not have learned to walk in a minute or so after birth. The baby was born with knowledge of walking, i.e., coordinated control over muscles of four legs and the need to follow its mother.
>
> Jim Spilker, a Stanford EE PhD, founded a Silicon Valley company named Stanford Telecommunications. This company had a contract with the US Air Force to develop the algorithms and software for the nation's GPS system. In order to make it work with great precision, he explained to me, it was necessary to take into account the effects of General Relativity. Earth's gravity has an effect on satellite radio propagation. Not taking this into account could result in errors of hundreds of feet in position at the surface of the earth.

A human baby must be born with much more knowledge. It seems likely that the face of a female was pre-recorded in memory. This innate image would be a generic face image passed down from one generation to the next. It would be close enough that the baby would be looking for a female face, finding "mama" and bonding to her. The actual mother's face would subsequently be recorded by baby over and over again in memory, cementing mother–child bonding.

With a pre-recorded image of a woman's breast, the baby would learn that on finding it, it provides nourishment. Fascination with women's breasts seems to be in-born and curiously, it persists for both males and females throughout life.

The baby is born with knowledge of how to walk. After muscle development, the baby becomes a toddler at about one year of age. Becoming a toddler takes place, with many falls, over a week or so. The baby could not have learned walking control in such a short time. This must have been there before the baby was born.

The baby is born with knowledge of how to use its voice to make loud noises, but is not born with knowledge of how to speak. That is a learned skill over a period of years after being born, and it is learned by reinforcement learning.

18.2 Storage and Retrieval of Innate Knowledge

All this learning and pre-born information is stored in long-term memory. It is conjectured that the memory storage and retrieval mechanism required for in-born knowledge is the same as for sensory knowledge learned during a lifetime.

18.3 DNA

At the moment of conception, a sperm and an egg unite to form a new cell containing deoxyribonucleic acid (DNA) from both parents. This cell will multiply and become a fetus. The single original cell contains the innate knowledge. This knowledge must be stored somehow and somewhere. The original cell nucleus contains this DNA. Innate knowledge must reside in this DNA. Where else could it be?

The mechanisms of storage and retrieval for in-born knowledge are the same as for knowledge obtained after birth. The innate knowledge is stored in DNA. As the baby matures, this knowledge remains stored in DNA. The mechanisms of storage and retrieval work with this DNA. New knowledge will also be stored in DNA. Access to innate knowledge and new knowledge is seamless. Knowledge is retrieved from both sources in the same way. This suggests that DNA is the storage medium for all knowledge.

DNA storage is digital. How new knowledge is transferred to DNA and how it is encoded are subjects for future research. The DNA of the skin of an adult, for example, contains the original innate knowledge. This does not change. DNA of the cells of the brain do change in order to record new knowledge. The DNA of the body is not totally immutable. Viral infection can change DNA.

The DNA of the original single cell was able to contain all of the innate knowledge. Imagine the amount of information that could be stored within the DNA of all the cells of the brain, neurons, ganglia, etc. The storage capacity for long-term memory is so vast that it is not necessary to erase knowledge to make space for new knowledge. Long-term storage of information is permanent.

Initially, all knowledge is stored in the nucleus of the first cell. There are no neurons, neural networks, or memory traces in this nucleus. Memory exists there without neurons, only DNA. All knowledge is stored in the brain's DNA, not in the memory's neural networks. The neural networks have the function of retrieval, not storage.

18.4 Manufactured DNA for Digital Data Storage

There is a great deal of interest in developing archival data storage methods based on manufactured (synthetic) DNA [1, 2]. To promote this approach, the DNA Data Storage Alliance was formed in 2020. There are dozens of members of this Alliance, including IBM, Microsoft, Battelle, Fujifilm, Los Alamos National Laboratory, Semiconductor Research Corporation, Seagate, etc. [3].

In June 2021, the Alliance published a paper "Preserving our Digital Legacy: An Introduction to DNA Data Storage" [2]. This paper summarizes the state of knowledge at that time and makes future projections for DNA data storage.

Here is a quote from the Data Storage Alliance [2]:

> There are phases in the lifecycle of technologies in which huge transformations lurk. We believe we are on the cusp of one such transformation today.

This is referring to archival storage using manufactured (synthetic) DNA as a storage medium.

Although this effort has yet to bear fruit, it suggests that the data in long-term animal memory can be stored in DNA.

18.5 Summary

Knowledge of how to walk, a walking algorithm, is inherited before birth. After a year of muscle development, the human baby is able to use the innate knowledge and begins to walk. The inherited walking system is generic, not tailored to the individual baby. Initial walking is awkward until adaptation takes place in the motor cortex. Adaptation is effected in an unsupervised manner by means of a reinforcement learning algorithm. Over time, the adaptive algorithm maintains a fine gait, even as physical changes take place. Innate knowledge is a very complex subject. There are many examples of in-born knowledge. They are designed to give the newborn a head start in life and a means of survival.

Appendix

Use google (or your favorite internet search engine) to explore:

- Innate knowledge
- Innate knowledge stored in DNA
- Innate knowledge for walking.

References

1. Landsman, D., Strauss, K.: The DNA data storage model. Computer **56**(07), 78–85 (2023). https://doi.org/10.1109/MC.2023.3272188, https://doi.ieeecomputersociety.org/10.1109/MC.2023.3272188
2. DNA Data Storage Alliance: Preserving our digital legacy: an introduction to DNA data storage. https://dnastoragealliance.org/dev/wp-content/uploads/2021/06/DNA-Data-Storage-Alliance-An-Introduction-to-DNA-Data-Storage.pdf
3. Data Storage Alliance. www.dnastoragealliance.org

Chapter 19
DNA and Memory

Abstract At conception, a new cell with DNA from both parents is created. The DNA contains the formula needed to construct the animal's body and to construct the animal's brain. The brain is constructed with all innate information pre-loaded. Information to be stored long-term is stored in the brain's DNA. How data, images, patterns, and sounds are encoded in the brain's DNA is unknown.

At the moment of conception, DNA is taken from the mother and father to form a new cell and is the start of a new living animal.

19.1 Innate Storage in DNA

The new cell DNA contains the information ("blueprint") needed to construct the entire living animal. Also contained in the DNA is the inborn information that will be "pre-loaded" into the developing brain. This innate, inborn information is derived from both parents and could be stored nowhere else except in the DNA in the single original cell. There are no neurons and synapses in the single cell. There are no memory traces. Yet complex information is being held. Held where? DNA is the only logical answer.

It is believed that the mechanism for long-term storage and retrieval of the information gained during a lifetime is the same as that of storage and retrieval for innate information.

19.2 Long-Term Memory Stored in DNA

The conclusion is that long-term memory is stored digitally in the brain's DNA, or possibly RNA. It is not stored in the neurons and synapses of the brain. The wiring of the brain is random, not controlled by innate knowledge.

B. Widrow and E. P. Katz, *Cognitive Memory*, Springer Series on Bio- and Neurosystems 17, https://doi.org/10.1007/978-3-031-80939-2_19

19.3 Role of Brain's Neural Networks in Memory

Alzheimer teaches us that the brain's neural networks are certainly involved with memory. The question is,

> What role do neural networks play regarding memory function?

It is believed that the neural networks, not storing long-term memory, perform the critical function of memory retrieval.

19.4 Summary

Major concepts:

- All information stored in the brain's memory is stored in DNA or RNA as digital memory.
- The memory mechanisms for storing innate knowledge is the same as that for long-term storage of knowledge accumulated in a lifetime.
- Stored information is not ultimately stored in the neurons and the synapses. The neurons and the synapses do play a key role in association and retrieval of stored information, but not in the actual memory storage.

Appendix

Use google (or your favorite internet search engine) to explore:

- DNA computer
- DNA computer Tic-Tac-Toe
- Knowledge in DNA at conception
- Knowledge stored long-term in DNA

Chapter 20
Prompting the Memory

Abstract When attempting to recall "something" from long-term memory, the effort is focused on finding the prompt. Upon acquiring the prompt, recollection is immediate. Without a prompt, long-term recollection is not possible. Short-term memory records current patterns and images, interesting or not. Everything is recorded. Retrieval is based on a sense of timing and a prompt.

20.1 Searching for the Prompt

When one knows "something," but cannot recall it when needed, this is frustrating indeed. Knowing "something" implies that it is stored in long-term memory. The reason why the "something" cannot be recalled at a particular time is that the correct prompt is not available. Given the correct prompt, the "something" is recalled almost immediately.

To accomplish fulfillment of the mental process for finding *something*, one is really searching the environment for the correct prompt. Figuratively *scratching one's head* to recall *something* is actually invoking several different memory heuristics to discover the prompt.

A younger person simply says, "I can't remember." when unable to complete a search for *something*. When an elderly person cannot complete a search, that person says "I am having a senior moment." In both cases, the failed search by a young person is the same phenomenon as a senior moment by an elderly person. The fundamental problem is finding the prompt.

20.2 Recalling a Name

In Chap. 1, there is a story about name recollection. Both Mrs. Widrow and Professor Widrow were trying to remember the name of a well-known friend to both. Coincidentally, there were two concurrent senior moments. This seemed silly. How could

B. Widrow and E. P. Katz, *Cognitive Memory*, Springer Series on Bio- and Neurosystems 17, https://doi.org/10.1007/978-3-031-80939-2_20

the name of a well-known person be forgotten, especially by two people at the same time?

The next day, the friend's illusive first name was recalled by Professor Widrow. But he could not remember the last name. He told Mrs. Widrow the first name, and she immediately recited the last name. Something prompted the first name for Professor Widrow after festering in his mind for about a day. The first name served as a prompt to Mrs. Widrow to recall the last name. Given the full name, it became possible for both of them to recollect all sorts of occurrences with that friend.

20.3 Visiting with a Psychiatrist

A psychiatric patient is placed in a comfortable position (on a "Freud's couch") in order to have a question-and-answer session with the psychiatrist. Something is stored in the patient's long-term memory that is very troubling. Neither the patient nor the psychiatrist knows what this is, and they are trying to discover it. The problem is that they do not know the prompt. After many sessions of back-and-forth discussions, the prompt finally appears. When this happens, the patient is then clearly able to recall details of the recovered memory.

20.4 Schizophrenic Hallucinations

A person with schizophrenia has hallucinations during which vivid scenes and conversations appear, though dreamlike. The patient experiences recollections that were not requested and spontaneously come from memory without prompting.

20.5 Memory and Alzheimer's Disease

There are many known symptoms that are associated with Alzheimer's disease. One type of symptom is the following:

> A lady living in a neighborly house had developed Alzheimer symptoms. One could have a fine conversation with her and not think that she has Alzheimer's disease. But talking with her husband yields a different story. She would ask him a question, and he would give an answer. About 15 minutes later, she would ask the same question again, and he would give her the same answer. This repeats over many cycles. What was happening was that her environment provided her with the same prompt repeatedly, and this would cause her to re-ask the same question. She did not have any memory of a received answer due to her failing short-time memory. Her very short-term memory, 10 second memory, was working, so that she could carry on a conversation. Interestingly, she could recall data recorded long ago. For example, she could remember how to speak English. But she could not recall data and images that were just recently presented to her. She died from Alzheimer's.

How could that happen? Her heart and lungs were working well. She had become bedridden because the disease had attacked that portion of her brain which generates signals for locomotion and motor control. Being bedridden full-time seems to have fostered the onset of pneumonia, and that proved to be fatal.

20.6 Memory of Dreams

When falling asleep, the mind goes through a twilight phase as a transformation goes from being awake to being asleep. When waking up in the morning, the brain once again goes through a similar twilight phase, half asleep and half awake. These waking twilight times should not be dismissed out of hand because they are very important opportunities for thinking and creativity. Most people are unaware of this phenomenon and their potential opportunities.

If one tries to solve a problem when going to sleep, short-term memory is involved. The problem itself can become a prompt and going into twilight carries the prompt into sleep. An attempt to find a solution often continues into the sleep dream cycles. Everything thought goes into short-term memory, and if important, into long-term memory. Recollection of the dream is possible with a suitable prompt.

Sometimes one awakens in the middle of a dream. The exact details of the dream could be recited. Obviously because the dream is stored in memory. When awakened in the morning, any dreaming during twilight time (the last REM cycle), the dream's details can be recalled. If a solution to the problem or an approach to a solution is uncovered during the dream, this can be preserved if written down immediately upon waking up. Otherwise, it will most likely be forgotten, and the dream will be lost.

> One night several years ago, Prof. Widrow awakened at 4:00 am during the middle of a dream. His dream had presented a solution for a control system problem that he had been trying to solve for over ten years. He knew that his memory had a solution and immediately wrote it down. Then he fell back into a deep sleep, very satisfied and content. The next morning, he called Eugene Walach, his postdoctoral scholar and described the solution. Eugene simulated this on a computer, but it did not work. By moving one simulated wire to another side of a circuit summation point, Eugene then had a simulated system that worked as expected. They were so excited that they decided to write a book. [1]

When falling asleep and thinking of nothing significant, you will have dreams during REM sleep. About every 90 min, you will undergo a sleep cycle that includes a dream phase. These dreams are for the most part unprompted, it seems. If, when waking in the morning and you were not dreaming during twilight time, you would be totally unaware that you had been dreaming during the night.

On August 20, 2022, CNN broadcasted an interview by Fareed Zakaria with American singer, pianist, and composer Billy Joel. During the interview, Billy Joel described his sometimes, nighttime experiences, waking with a new tune in his thoughts. He had a pen and pad of paper by his bedside and wrote down his thoughts immediately. He placed a tape recorder at this bedside so he could sing his latest creation, but this did not always work. He wrote, but did not sing. Why? Who knows?

Maybe he did not want to disturb his wife? He would not be able to remember anything unless he awoke in the middle of a dream. Otherwise, there would be no prompt and his ingenious thoughts would be lost.

20.7 PTSD

PTSD (post-traumatic stress disorder) can manifest in different ways, variations due to the nature of the traumatic experience and the reaction of the individual. Generally, memory of the traumatic event is experienced at night in the form of nightmares. In the morning, if waking during a nightmare, the memory of the traumatic event would be retrieved and stored in short-term memory. By repeating over and over again, it could be recalled during daytime in the form of unprompted recollections.

A soldier returning home after experiencing combat may be troubled enough to seek the help of a therapist. The soldier can describe a horrible experience. There is apparently another experience even more horrible and more troubling that does not surface. The soldier and the therapist would like to discuss this deeper experience. After many sessions of therapy, they were finally able to do this. The expectation was that the discussion would provide at least some relief from the symptoms.

For this particular case, one can only speculate about how all this works, and what is the role of memory. It would seem that both traumatic events were recorded in long-term memory being sufficiently "interesting." Without prompts, they would not be recalled except for the possibility of retrieval without prompting during sleep and dreaming, causing nightmares. Awakening during a nightmare, a vivid recollection could be captured and recorded in short-term memory. Repeating over and over again, this experience could be preserved and discussed with the therapist. The second experience will cause occasional nightmares, but will remain below the surface if and/or until caught during a nightmare, and transferred to short-term memory, or if the soldier and the therapist discover the prompt for direct retrieval from long-term memory. With vivid recall and ensuing discussion, some symptom relief results.

This is only one case, and the memory action described above is pure speculation. The soldier and his symptoms have been fabricated, derived from about a half-dozen actual clinical cases reported in an article in the New York Times titled "Study Suggests Trauma Stays in Present Tense," by Ellen Barry. The reference is New York Times, Tuesday, December 5, 2023, page D-3. A soldier and his symptoms are not represented exactly by any of the reported clinical cases. This case was fabricated to illustrate how one may go about analyzing an actual clinical case.

20.8 Summary

When symptoms of Alzheimer's disease manifest, considerable destruction of the memory's synapses and neurons has taken place. The resulting dementia is debilitating. Short-term memory is attacked, preventing the recording of new data. Old data is still accessible from long-term memory if prompted correctly. Finding the prompt is problematic for demented patients. On a bad day, certain sought after data cannot be recalled. On a good day, the same data will be recalled from long-term memory. Prompting could come from the environment.

Data retrieval is based on prompting. Psychiatric treatment, talk therapy, is all about seeking the unknown prompt, the key to evil thoughts stored in long-term memory. Finding the prompt could sometimes take years of weekly talk sessions.

Appendix

Use google (or your favorite internet search engine) to explore:

- Dreams and memory
- Alzheimer's disease
- Schizophrenia
- Freudian psychotherapy
- Long-term memory prompting
- Short-term memory prompting
- Dreams and memory
- Alzheimer's disease
- Schizophrenia
- PTSD.

Reference

1. Widrow, B., Walach, E.: Adaptive Inverse Control. Prentice Hall (1996)

Chapter 21
Long-Term and Short-Term Memory

Abstract Input data is stored permanently in long-term memory, if it is "interesting." Whether it's interesting or not, all input data is stored temporarily in short-term memory. The same segment designs for long-term memory by nature are also incorporated into the short-term memory segments. Long-term memory records a database that accumulates over a person's lifetime and includes innate data (Camina and Guell in Front Pharmacol 8:e00438, 2017). Input data is seen through a moving window in short-term memory. The window width (its "span") determines the amount of data stored at any moment in time. The short-term memory consists of a large number of memory "elements," each with its own timespan. Some elements may have a very short span, on the order of 10 s. Others may have medium-length spans, on the order of one hour. Yet others may have much longer spans, on the order of a day, possibly longer. With an intuitive sense of time, a volitional selector switch determines the element to provide the short-time memory output. Prompt signals trigger recall for both short-term and long-term memories.

Thus far, our focus has been on long-term memory. Everyone has had experience living with long-term and short-term memory. But why have both? Could all memory functions be accomplished with long-term memory alone?

21.1 Short-Term Memory

This chapter develops a short-term memory design. It is highly speculative to assume that this is in fact nature's design. But the design proposed in this chapter serves as a reference model for human short-term memory. This model is designed to exhibit human-like memory characteristics, in accordance with living experience.

The basic ideas for long-term memory are expressed in Figs. 8.1 and 8.2. The basic ideas for short-term memory are hinted at in Fig. 13.1. The purpose of this chapter is to study short-term memory in depth, and its integration with long-term memory.

B. Widrow and E. P. Katz, *Cognitive Memory*, Springer Series on Bio- and Neurosystems 17, https://doi.org/10.1007/978-3-031-80939-2_21

Streaming input data and patterns are stored permanently if they are selected using the "interesting" switch. Long-term memory records only a limited set of input data.

Short-term memory does not have an "interesting" switch. Streaming input data goes directly to memory "elements." Nothing is stored permanently. As new input data arrives, it is stored for a finite time temporarily with the oldest data being deleted.

21.2 Short-Term Memory Elements

It seems that nature has provided us with many types of short-term memory. For example, a very short short-term memory element holds input data for perhaps ten seconds. A medium length memory element holds input data for perhaps an hour. A longer length memory element holds input data for perhaps as much as a day or longer.

Evidence for a very-short length memory element is, for example, needing to repeat a telephone number again and again, as long as it's needed. Without the repetition, this number could be remembered for about 10–30 s. Evidence for a medium length memory element is suggested from reading a newspaper paragraph. What did it say? You cannot recall the actual words, but you can recall a mental picture derived from the text. If the paragraph is highly interesting, it will be recorded in long-term memory and temporarily stored in short-term memory. If the paragraph's essence is not stored in long-term memory, the paragraph can be recalled over perhaps 1 h. Evidence for a long-length memory element of the order of a day or less is apparent when one can recall events over the last day. When recalling brushing your teeth before bed, was this done today? Or was it yesterday's brushing also being visualized? A memory length of less than a day assures that the visualized fact of brushing actually occurred today.

21.3 Remembering Procedures

The procedure for teeth brushing is stored in long-term memory. When engaging in the teeth brushing procedure, one does not recall the details. The brushing procedure is automatic, essentially acting "on autopilot." Only the fact of the brushing action is stored, temporarily, in short-term memory. The details were never recorded.

21.4 Graphic Representation of Long-Term and Short-Term Memories

The basic idea of long-term memory is expressed in Fig. 21.1. Each rectangular box represents a memory segment of the type seen in Figs. 8.1 and 8.2. The numbers on the boxes represent specific blocks of data where the size of the data blocks is commensurate with the memory segment size. Of the segments shown in Fig. 21.1, some have recorded data, and some are empty, waiting to be filled. Figure 21.1a represents recorded data at a particular snapshot in time. As indicated, data was recorded prior to the recording of data (at time step-1). As shown, data has streamed in from the memory input line and filled segments with data blocks (– 1), (0), (1), and (2).

Figure 21.1b shows the recordings one unit of time later. Data block (3) has arrived and has been stored

Figure 21.1c shows the state of the recordings one time unit later. Data block (4) has arrived and is stored into the next empty segment. The data in each segment is stored permanently. Empty segments are filled as more data arrives on the memory input line. A unit of time is defined as the time elapsed while filling the memory of a single segment.

Figure 21.2 illustrates the working of the short-term memory process. Data enters via the short-term memory input line. All data is recorded without discrimination. There is no "interesting" switch, and the data is recorded for temporary use. Data is retrieved from this memory in response to prompt inputs that arrive on the prompt input line (not shown in Fig. 21.2). These prompts arrive directly from sensory inputs without being selected by an "interesting" switch. Both long-term and short-term memory systems are composed of the same basic parts. A fundamental difference between long-term and short-term memory is the "interesting" switch. Long-term memory stores input data permanently and short-term memory stores everything temporarily.

Figure 21.2a illustrates data storage at a particular snapshot in time. Data blocks (1), (2), (3), (4), and (5) have been recorded. One time unit later, Fig. 21.2b shows

Fig. 21.1 Elements of long-term memory

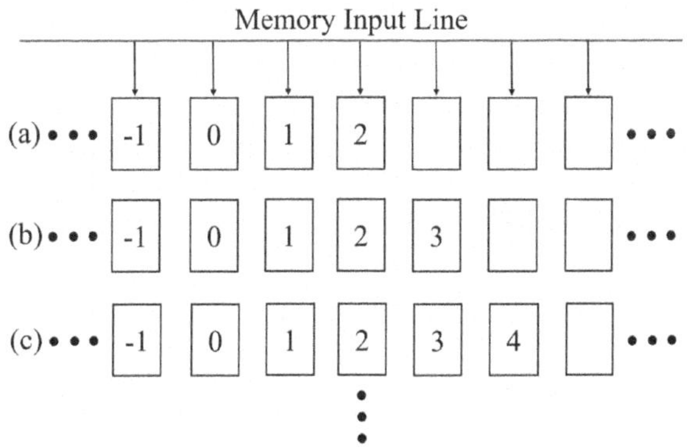

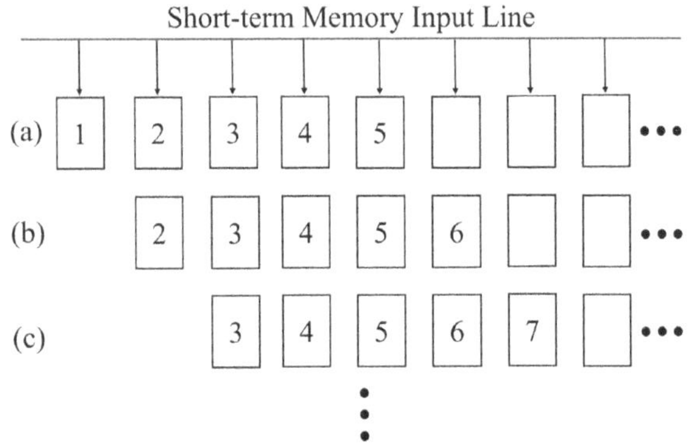

Fig. 21.2 Elements of short-term memory

the recorded data, except that data block (1) has been deleted and newly arrived data block (6) has been recorded into the next available segment. Figure 21.2c shows the state of the recordings at the next later time unit. Data block (2) has been deleted and new data (7) has been recorded. Figure 21.2 illustrates a short-term memory functioning with a span of five-time units. At any given time, this memory is capable of storing five blocks of data in five segments. This memory deletes and stores, deletes and stores, etc. The memory behaves in a manner reminiscent of an inchworm.

Returning to the question of why does nature provide us with both long-term memory and short-term memory? Data stored in long-term memory is highly filtered by the "interesting" switch. With most input data being uninteresting, only a small fraction of sensory data will be stored. Storage space is essentially unlimited, but there is an advantage from limiting the amount of data permanently stored. Less data would make retrieval easier, with far fewer false hits.

Short-term memory has elements with long spans on the order of a day, and elements with shorter spans on the order of 10–30 s. There can be many more elements with intermediate spans.

Figure 21.3 contrasts the structures of long-term and short-term memories. Figure 21.3a shows a long-term configuration with an "interesting" switch. Figure 21.3b shows a short-term memory with multiple elements, three of them as illustrated. One element has a span of three. Another element has a span of two. A third element has a span of one.

The memory elements of Fig. 21.3 are all recorded simultaneously. A time selector switch determines which element is to be selected to produce the short-term memory output. The selection is volitional, based on an intuitive sense of time. Did the event occur a few seconds ago? Or did it occur within an hour ago? Or sometime today? "What did I have for dinner last evening?".

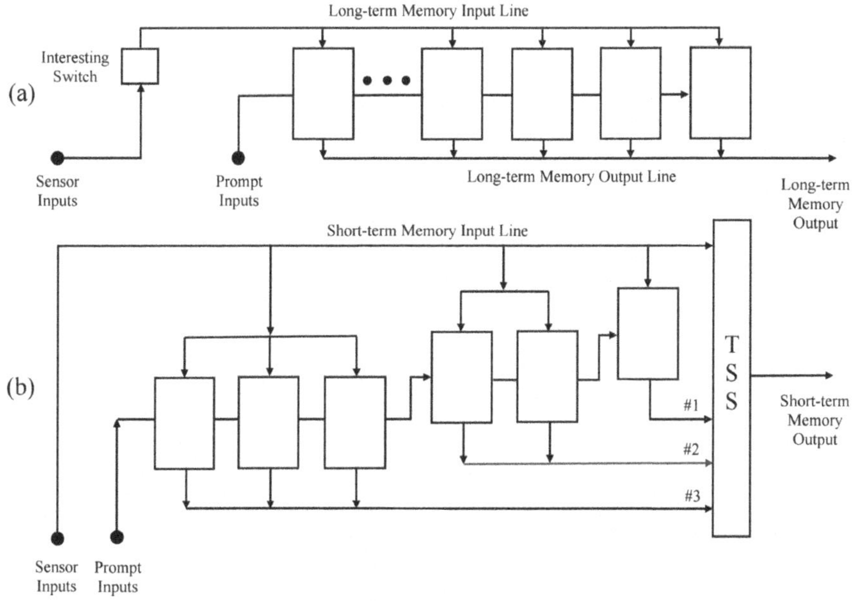

Fig. 21.3 **a** Long-term cognitive memory. **b** Short-term cognitive memory. *TSS* Time Selector Switch

21.5 Summary

Both long-term memory and short-term memory are active and "turned on" simultaneously. Long-term memory records only if input data is deemed to be "interesting." Short-term memory records all input data, interesting or not [1]. Recordings of long-term memory are permanent. Recordings of short-term memory are temporary, obtained by scanning over the streaming input data using a moving window. The width of the window is its span. Short-term memory elements with a variety of spans scan the streaming input, moving over the input data like an inchworm. An intuitive sense of time, approximately when an event happened, is used by an individual to select the element with an appropriate time span. The selected element provides the short-term memory output when triggered by a prompt input.

Appendix

Use google (or your favorite internet search engine) to explore:

- Long-term memory
- Short-term memory

Reference

1. Camina, E., Guell, F.: The neuroanatomical, neurophysiological and psychological basis of memory: current models and their origins. Front. Pharmacol. **8**, e00438 (2017)

Chapter 22
Dementia

Abstract Alzheimer's disease is the most common cause of dementia. Tangling of the Tau protein within the brain is the primary cause of the disease. The tangles sequentially attack portions of the brain as the disease progresses with synapses being especially vulnerable. The attack is debilitating to the neural networks and ultimately to the human subjects. Destruction of the neural networks occurs simultaneously in both long-term and short-term memory. This disease is incurable, but promising approaches to the cure are in development.

Alzheimer's disease [3] is the most common form of dementia [2]. It is caused by abnormal buildup of amyloid and Tau proteins in the brain. Excessive amyloid causes "plaques." Excessive Tau causes "tangles" in brain cells. Plaques and tangles are destructive to neurons, synapses, glia, etc.

22.1 Destruction of Neural Networks

Every segment of long-term and short-term memory employs neural networks for data retrieval, subject to prompting. With the onset for gradual destruction of neurons, the networks continue to function properly as long as the networks are constantly trained and re-trained. Each segment does this for itself. Symptoms develop when the neuronal destruction advances to the extent that even re-training will not preserve correct functioning. Network outputs become erroneous, unreliable, and gradually produce more and more wrong answers. In some cases, neuronal destruction can begin as early as 20 years before symptoms appear.

B. Widrow and E. P. Katz, *Cognitive Memory*, Springer Series on Bio- and Neurosystems 17, https://doi.org/10.1007/978-3-031-80939-2_22

22.2 Short-Term, Long-Term Memory Loss

There is a widely held belief that Alzheimer symptoms begin first with the loss of short-term memory, followed sometime later by loss of long-term memory. The notion that all forms of memory are attacked simultaneously we believe is correct. All memory segments have neural networks. The plaques and tangles do not distinguish between long-term neural networks and short-term neural networks. They simply attack any kind of neural network. They attack networks in segments containing recorded data as well as in segments that are empty of data.

Logic and reasoning tell us that memory loss occurs at the same rate in long-term memory as in short-term memory. If that is the case, why does it appear that short-term memory loss occurs prior to long-term memory loss?

In the case of long-term memory, destruction of a fraction of all its segments makes certain items of data inaccessible. This will not be realized until prompting fails to make a hit. Under this circumstance, it's unknown if the desired data does not appear because of memory loss or that the data was never recorded at all in the first place. If the prompt hits a segment that has not yet been destroyed, ancient data stored in memory is retrieved, and the person believes that the long-term memory is functioning correctly. But in reality, the observed functional decline has already begun.

The situation with short-term memory is quite different. Destruction of a fraction of the segments results in no hits when prompting them for specific data. The person knows it should be there. But if the specific data was recorded in a deteriorated segment, the person would sense this immediately when seeking retrieval and would think that this is a "senior moment" or something else. If this persists, there is a short-term memory problem. The person would be aware of this long before becoming aware of loss of long-term memory.

Alzheimer's disease begins with loss of short-term memory. This perception is not reality. Both forms of memory degradation occur simultaneously.

22.3 Examples of Short-Term Memory Troubles

- Trouble remembering telephone numbers.
- Confusion in going room to room. "What am I doing here?"
- Brushing teeth. There are four quadrants: lower left, lower right, upper left, and upper right. When brushing the upper right, have I already brushed the lower left?
- Dinner last evening. What was served?

22.4 Long-Term Memory Troubles

- Death of segments destroys some recorded data. New data can still be stored if stored in intact segments.
- Data recorded and subsequently destroyed by the disease is often not missed as there is so much of it.
- Long-term memory still functions with partial loss of some data, but with ever growing amount of loss, there will not be much left of long-term memory.

22.5 Postscript

In a recent publication of the NIH Research Matters [1], some of the latest research results about Alzheimer's disease are reported. The major driver of Alzheimer's is the Tau protein [4]. The buildup of Tau and its tangles appears to attack the synapses that are used for neuron to neuron signaling. Loss of synapses would have the same effect as loss of neurons. Researchers believe that the gradual decrease in synaptic density causes the progressive symptoms of the disease.

22.6 Summary

Alzheimer's disease is primarily caused by the tangles of the brain's Tau protein. This is a gradual process that generally extends over the years. The tangles attack synapses that are neuron to neuron connection points in the neural networks of long-term and short-term memory segments. Within a given segment, as the synapses are slowly destroyed, the neural network continues to function since it is constantly re-trained. This is short-lived until the continued destruction eliminates enough synapses to degrade correct functionality and ultimately renders the neural network ineffective. The segment's network fails and with it, the segment and its memory content become inactivated as the disease progresses. There is no cure for Alzheimer's disease, but experiments with rats show promise of a solution.

Appendix

Use google (or your favorite internet search engine) to explore:

- Dementia
- Alzheimer's disease (Many papers discuss causes and symptoms)

References

1. NIH Research Matters: NIH (National Institutes of Health publishes this report that describes the latest research on the disease, its causes and developing cures) (December 20, 2022)
2. Wikipedia: Dementia
3. Wikipedia: Alzheimer's disease
4. Wikipedia: Tau protein

Chapter 23
The Use of Memory: The "Executive"

Abstract In this chapter, we postulate the existence of a portion of the brain which incites executive action in response to the cognitive memory output. The executive's location in the brain is unknown; perhaps it is a part of the cerebral cortex? The memory's output signal becomes the executive's input signal.

Previous discussion has been primarily concerned with the various mechanisms that nature uses to derive the memory output signal. What is done with this signal is the subject of this chapter.

23.1 The Memory "Customer"

Question: *What part of the brain is the "customer" for the brain's memory output signal?*
What does the customer do with this signal?

The part of the brain that consumes this signal is an enigma. Its location within the brain is unknown, and its function is not fully understood. We can only observe human and animal behavior, and from these observations, an inference can be created of the function and construction of the memory's customer [1].

23.2 The "Executive"

When information is derived from memory, something is generally done with it. Executive action takes place in response to the memory output. We shall name this responsible part of the brain the "executive." Like the memory system, the executive makes primary use of neural networks [2–4]. While the executive's structure is unknown, it is highly likely that it was developed in early gestation well before birth. One can only speculate about this.

B. Widrow and E. P. Katz, *Cognitive Memory*, Springer Series on Bio- and Neurosystems 17, https://doi.org/10.1007/978-3-031-80939-2_23

23.3 Examples of Executive Actions

It was once believed that fish have no fear, and that a hook in the mouth does not cause pain. This is nonsense. Pattern recognition is critical for survival. When a situation is recognized as dangerous or painful, the executive takes action. When a fish recognizes an unusual object nearby and coming closer, the fish is frightened and swims away at full speed. The swimming away response is triggered by executive action, in response to patterns of danger recognized by the fish's cognitive memory. Fish do have memory. Some fish spawned in a particular river spend most of their lives in the ocean. Toward the end of life, they return to the very same river and very spot where they were spawned.

Watching a small lizard from a distance sunning itself on a sidewalk, a person coming near obviously terrifies the lizard. It recognizes danger and remembers which direction to go to reach the shelter of the bushes overhanging part of the sidewalk. This little animal runs at full speed seeking protective shelter. The action of running away is an executive response to recognized fearful threats sensed by cognitive memory. This could not be done without memory. Flies behave like this, mosquitoes behave like this, all animals behave like this. They all take executive action for survival.

23.4 Summary

In the human brain, the executive function might be performed by a part of the cerebral cortex. The function and structure of the executive, evidenced for example by the above animal observations, are not fully understood, nor is its location within the brain known. The executive is most likely constructed of neural networks.

If one wished to construct an artificial executive with neural networks, deep analysis would be required to create a robust artificial structure that would exhibit at least some of the executive functions. This has yet to be accomplished because primarily the executive is enshrouded in unknowns, the only aspect of it that seems realistic is that it exists.

Appendix

Use google (or your favorite internet search engine) to explore:

- The executive, human memory
- Memory storage in the human brain
- Memory processes in the human brain.

References

1. Goldenberg, G.: Neuropsychology of memory and the central executive. In: Henn, F., Sartorius, N., Helmchen, H., Lauter, H. (eds.) Contemporary Psychiatry. Springer, Berlin (2001). https://doi.org/10.1007/978-3-642-59519-6_13
2. Andres, P.: Frontal cortex as the central executive of working memory: time to revise our view. Cortex **39**(4–5), 871–895 (2003). https://doi.org/10.1016/S0010-9452(08)70868-2
3. Executive Functions: Weill Institute, University of California San Francisco (2023). https://fatiguemanagersnetwork.org/wp-content/uploads/Executive-Functions_Weill-Institute-for-Neurosciences.pdf
4. Wikipedia: Executive functions

Part V
Cognitive Memory and Thinking

Thinking by our earliest human ancestors was focused on finding food and water, finding shelter from the elements and from predators, and finding a mate. This is similar to the evident thinking of all animals, from spiders to mountain lions. Over the millenia, human thinking evolved as objects of the environment changed. Evolution brought primitive thinking to the thinking of today, designing cell phones and rocket ships for the exploration of Mars.

Cognitive memory plays a key role in all forms of thinking. Memory allows the creation of thinking threads from moment to moment. Without memory, discoveries will be forgotten, and forward evolution would not be possible.

Associated with thinking is the extraction of meaning from a scene or from a thought. Meaning is established when a new object's properties, prompting a cognitive memory, have a hit with previously recorded images. The contents of the hit folder, previously studied and evaluated, transfer to the new object and shed knowledge upon it.

Extraction of meaning of a scene or a thought involves creation of a mental image, a simplification, and a compression and association of this with previously stored mental images.

Information storage in long-term memory can be highly detailed, as with eidetic memory. This is rare and exceptional. In most cases, except for rote learning, only mental images are stored in long-term memory. Scenes from long ago can be remembered as mental images and made sense of by relating them to data previously stored in cognitive memory.

There is a question. With only sensory inputs and a cognitive memory, would that be sufficient to realize an elementary thinking process? What else might be needed? These questions are in the background of Chaps. 24 and 25.

Chapter 24
Thinking

Abstract A major "application" of cognitive memory involves the act of thinking. Sensory inputs are processed, clustered, and recognized by cognitive memory, with the results being supplied to an executive for purposeful execution. The result is situational improvement for the thinking animal. A "thinking system" consists of (1) sensors (eyes, ears, touch, etc.), (2) a cognitive memory, and (3) an executive to implement the cognitive memory output. Nature's purpose for the thinking system is to improve the circumstances of the thinking animal. Thinking began in animals striving to satisfy their basic needs of food, water, shelter, reproduction, etc. Satisfaction of these needs was and still is essential for survival of the species and requires elementary thought. Our human ancestors were hunter-gatherers, and their thinking revolved around objects found in their surroundings. They began to alter their natural environment, and this led to the development of tools to aid in hunting (and much later to agriculture), building shelters, etc. The more that they altered their world, the more objects they had to think about. One step leads to another. Thoughts build upon thoughts, and memory was essential to sustain these thoughts. They were able to produce sounds and eventually began to devise sounds that corresponded to objects, and eventually, language was born. Association of spoken "words" to objects made it possible to record thoughts in cognitive memory. Over eons of time, one step led to another, and now there are many other modern tools and conveniences which are outgrowths of thinking about objects and speech development.

24.1 What is Thinking?

Thinking started with the most primitive animals, perhaps as much as five hundred million years ago. As the animals evolved, so did thinking.

But, what is thinking? A dictionary definition that is most appropriate to describe the thinking process through all stages of its evolution is the following:

> Thinking, also known as 'cognition,' refers to the ability to process information, to store and retrieve memories, and to select appropriate responses and actions.
>
> *Ref:* University of Cambridge, Inclusive Design Toolkit [1]

B. Widrow and E. P. Katz, *Cognitive Memory*, Springer Series on Bio- and Neurosystems 17, https://doi.org/10.1007/978-3-031-80939-2_24

A human "thinking system" contains (1) sensory inputs (from eyes, ears, olfactory, etc.), (2) a cognitive memory to receive and process the sensory inputs, and (3) an executive to implement the thoughts stored in the memory. The purpose of this system is to think about the existing environment in order to modify it so as to create a new environment which is more advantageous to the human.

The same kind of thinking system exists within the most primitive of animals for those that have brains or at least some form of nervous system. These systems have evolved into the present day and are continuing to evolve.

24.2 Eric Kandel's Aplysia Experiments

An example of thinking in a primitive animal, Aplysia, is the following: Aplysia is a sea slug (mollusk) that has a primitive nervous system. It has sensory inputs that feed signals to a cognitive memory, which in turn, feeds signals to an executive.

Aplysia was used by Eric Kandel for neuroscientific research. Aplysia has large neurons, up to 1 mm. in diameter, that are relatively easy to study. It's a simple nervous system which contains about 10,000 neurons and has a primitive brain. This animal is not the earliest one that has a nervous system. It came to modern scientific attention as a result of studies by Eric Kandel, begun in the laboratory of Ladislav Tauc. Tauc had been studying the nervous system of Aplysia.

24.3 Eric Kandel

Eric Kandel is a Professor of Biochemistry and Biophysics at the College of Physicians and Surgeons of Columbia University. Professor Kandel was awarded the Nobel Prize for Physiology or Medicine in 2000 for discovering the role that synapses play in memory and learning. His experiments with Aplysia have become classics in Neurophysiology. The simplicity of the Aplysia nervous system allows one to understand the animal's entire behavior as related to the operation of individual neural circuits.

Professor Kandel was able to study synaptic plasticity with Aplysia and to relate synaptic change to learning. This phenomenon is common to all animals, including humans. One of the first to predict this on a theoretical basis was D. O. Hebb [3]. One of the first to observe this in a living animal was Eric Kandel.

Of interest is the gill response of Aplysia. Touch the gill, and it retracts. Electrically shock it, and it retracts. Sensory inputs come from touch or from an electric shock. Cognitive memory is formed of the Aplysia's neural network. The executive is also formed of the neural network. The executive controls the gill's withdrawal. The question is, "How does gill withdrawal improve the situation for Aplysia?" One can only speculate. It may have to do with survival. It may have to do with reproductive efficiency, or it may cause the animal to move to avoid a predator. Whatever the improvement is, it must be there or natural selection would not have caused Aplysia to be created with a gill withdrawal response. So, this is an example of thinking in a primitive animal.

Professor Kandel was doing experiments with training the neural networks of Aplysia. The gill response to a singular electric shock was noted. Repeated shocks caused synaptic change that resulted in reduced gill withdrawal responses to the shocks. This trained condition lasts about 2–3 weeks. In accord with all fundamental learning algorithms, a repeated stimulus causes sensitization and habituation. This was observed with Aplysia.

24.4 The Training Algorithm

The training algorithm that nature employs in the training of Aplysia is believed to be the Hebbian-LMS algorithm. This algorithm was developed by B. Widrow and is described in the recent book "*Cybernetics 2.0: A General Theory of Adaptivity and Homeostasis in the Brain and in the Body*" [2]. The Hebbian-LMS algorithm predicts habituation, and if the repeated stimulation stops, then there is overshoot and withdrawal.

When training an Aplysia, Professor Kandel was training an actual living, trainable neural network. What he called "memory" was the weight values of the trained neural network. No doubt, this is a form of memory. But it is quite different from the more comprehensive cognitive memory concept.

24.5 A Prof. Widrow Story

About ten years ago, I had the pleasure of meeting Eric Kandel. My wife and I were going to San Francisco for a symphony concert. I had heard that that afternoon, Eric Kandel was giving a lecture at UCSF. We were early arrivals at the lecture hall. Dr. Kandel was also early. He was at the front stage, checking the microphone and the slide projection equipment. And he then came down to the audience where there were a few early attendees. We were seated on an aisle. He came up the aisle and wanted to chat. He looked at me and said that I seemed to be someone that he had met previously. I had never seen him before. I must have looked like a cousin of his or some other relative. We chatted. I would have liked to talk to him about cognitive memory, but there was not enough time before his lecture on memory. I knew that would not be a good idea to distract him before he was to give a major lecture. So, I never discussed my ideas about cognitive memory with him.

24.6 The Next Sections of This Chapter

The remaining sections of this chapter trace the development of thinking from the beginning when animals first developed nervous systems. Keep in mind that the thinking system of interest consists of sensors, a cognitive memory, and an executive. The purpose is to manipulate the environment and objects within it in such a way as to improve animal well-being, to improve survivability and reproducibility. This is done by sensing the environment, recognizing patterns in it by the cognitive memory, sending control signals derived from the memory to the executive, the executive causing modifications to the environment, and if the modifications are an improvement, sending this to the cognitive memory for update.

24.7 Evolution of the Thinking Process

Nature's most important application for cognitive memory by far is thinking. Human thought is very complicated indeed. Human creativity is just short of being miraculous. The thought mechanism and processes have evolved over eons of time. It probably began with rudimentary thought similar to the thought exhibited by non-human animals, such as birds, reptiles, insects, mammals, and other life forms. The common thread driving the animal thought processes for the essential primary needs of food, water, shelter, and predator avoidance followed by secondary needs of procreation (protecting the offspring), companionship, and group structure (herding/flocking) continue to evolve.

How did this capability develop so long ago? One can only speculate. It seems reasonable to speculate that the most rudimentary thought processes began with the first living creatures having brains.

Mathematics has theorems built upon previous theorems and so forth, all the way back to original axioms. Likewise, the thought process started with rudimentary thinking by rudimentary animals and grew more complex with one manner of thinking built on top of previous ways of thinking. The thinking process of modern humans has evolved, starting with ancient rudimentary thinking.

24.8 Some Elements of Human History

The lineage of modern-day humans split from the early hominid ancestors about six million years ago in Africa. There they evolved over the next four million years into walking upright and becoming large-brained, primitive, prototypical humans. As they evolved, so did their thinking evolve over the next few million years.

The beginnings of human-like animals appeared about 5 million years ago. These primitive "people" were thinking, and their thought process has evolved over eons of time into modern human thought.

Later, there were the Neanderthals and Denisovans who came up from Africa about a half million years ago and spread to Western Europe and to Asia. The Neanderthals went west and the Denisovans went east. Ancient DNA samples indicate that these were the ancestors of modern humans, with DNA of Denisovan closer to present-day human DNA. However, both are linked to modern humans.

Paleoanthropologists have discovered Neanderthal fossils [5] that indicate that they were stocky, chinless humans. They also made tools for hunting which they apparently put to good use. They buried their dead. They made jewelry from eagle talons.

Both the Denisovans and Neanderthals made stone tools and were hunter-gatherers. They were thinking about how to catch game. It was possible for humans to use stone tools because their hands had four fingers with an opposing thumb. They could hold things in their grasp. A key difference comes from the shape of the human hand. Grasping things allows implementation of executive action.

Agriculture came much later and required deeper thinking. The development of agriculture marked a sharp turn in human history. Farming began about 12,000 years ago in the Fertile Crescent, the land that today includes parts of Iraq, Syria, Lebanon, Israel, and Jordan. The first crops consisted of peas, lentils, and barley. Wild goats and wild oxen were domesticated and herded.

Around the same time, agriculture also developed in Egypt. The Nile River, and its dependable flooding and fertile soil allowed the Egyptians to build an empire of great agricultural wealth harvesting the grains of wheat and barley.

Agriculture is sometimes called the "Neolithic Revolution." Hunter-gatherer lifestyles were swept away to be replaced by permanent settlements and reliable food supplies. Cities and civilizations thrived as crops and animals were farmed to meet the growing food demand of the emerging civilizations.

Agriculture was a mature art in Egypt 3,300 years ago when the Israelites led by Moses made their exodus from bondage. In the Sinai desert they could make unleavened bread with flour and water baked in the sun. The flour came from Egypt. This flatbread called "matzo" is eaten worldwide by Jews to commemorate the escape from slavery.

The time of Leonardo Da Vinci (about 1500 AD) [4] was a benchmark of human history. People had developed methods of making war with the crossbow and cannon at a high degree of sophistication. Leonardo was one of the greatest artists of all time as well as a great engineer and inventor. His sponsors funded him to invent weapons of war though none proved to be practical. He was interested in mathematics, and Euclid's geometry (formulated in the fourth century BC) played a significant role in his artistic work. Aside from its ever-present warfare between the great Italian families, the time of Leonardo marked an era of great invention and creativity.

24.9 The Origins of Human Thinking

The origin of thought began with ancient people. Throughout all this history, human thinking evolved in a similar way that human life evolved. Ancient humans were capable of modern thinking, but their priorities were shaped and directed by survival needs and living in primitive environments. The level of thinking leapt ahead with the development of agriculture. Planning ahead, necessary for farming, was added to the repertoire of thought.

24.10 Development of Vocabulary and Language

Animals create sound to communicate with each other. These sounds have meaning to the animals. Birds sing. Humpback whales [6] using. Lions roar. Cats meow. Like all other animals humans can make sounds.

Imagine a small group of early humans living together. A discovered object could cause one person to make a sound, and to remember the sound. Handing the object to another person and reproducing the same sound, the second person will begin to use that sound to identify the object. If the sound is agreeable to the group, all would collectively use that sound to identify the object. For this group, a vocabulary word is born. If this group has contact with other neighboring groups, they could all adopt the sound for the object or similar objects. This is the beginning of a language. In like manner, the objects become identified by respective sounds associated with the objects and accepted by the human community [7].

Single words for single nouns could evolve into phrases. For example, a meowing cat could be hungry. One person could say to another, "cat want milk." This would be a form of thinking. Phrases could evolve into sentences, and grammar could eventually develop. The "rules" of grammar evolved from usage of words and phrases.

Clusters of people could form a community around a common language. Another community hundreds of miles away would develop another language. People from one community would not be able to communicate with people from other communities.

As language developed, human thought developed. Words became a matrix upon which to develop and express thoughts. Cognitive memory is essential to allow a language to develop. People need to remember the associations between words, objects, and simple thoughts. Fluency requires memory, accuracy and quickness.

24.11 Human Characteristics

Humans stand and walk erect. The human hand has four fingers and an opposing thumb. Humans and especially young humans have an innate curiosity. Like all other animals, humans have fight or flight responses. Humans have a need for food and water, a need for shelter, and a need for companionship and intimacy.

Humans are born into a world of wonders, a rich environment of grasses, plants, trees, rivers, lakes, and land in addition to other animals. All these contribute to a basis for thought.

24.12 Baby Thinking

A few months old human baby begins communicating by crying, then starts babbling and imitating sounds. Early learning is dependent on communicating with others in order to begin to understand the world, and to ask questions of "Why?" At four months old, a baby can roll over in order to reach a toy nearby. The baby babbles and laughs and imitates sounds. At six months, the baby can support its weight on two legs when standing and holding onto furniture. The baby responds to its own name and babbles some vowel and consonant sounds. At twelve months, the baby is now a toddler that cannot be left unattended. With its newfound mobility, the toddler can exercise its curiosity and get into all sorts of things nearby and sometimes into trouble. The toddler now begins to utter "mama" and "dada" and tries to say words. At eighteen months, the growing child can say three or four single words, and possibly two-word phrases. At two years of age, the child uses combinations of two or three words together, knows the names of people, and knows the names of body parts. The child's expanded development now allows it to follow simple directions and can point to names of objects or objects pictured in baby books. At four years of age, the child has now learned the basic rules of grammar. Communication is clear, with speaking in complete sentences. The child knows words like "in," "on," and "under" and can sing a song or recite a rhyme from memory.

At each stage of development, thinking becomes deeper. A lot of this is due to increased physical ability. And of course, the more that one knows, the faster will be the accumulation of further knowledge all stored in cognitive memory. More complex thinking comes naturally as there are more pieces of information stored in memory that can be pondered mentally.

A baby being able to grasp a toy and move it in order to observe it from various perspectives is fundamental to thinking. The human hand plays a major role in the development of thinking.

24.13 Thinking and Memory

Could thinking be done with cognitive memory alone? Since the rudimentary substance of thought is based originally on sensory inputs, vision, auditory, etc., it seems that sensory inputs are necessary for thought. An exception is thinking during REM sleep, but all the images and sounds originally came from sensory inputs while awake. To have thinking, sensory inputs together with cognitive memory are necessary.

24.14 Modern Thinking

In this chapter, rudimentary thinking was all that was considered. The question is,

> How did this lead to modern mathematics, arguably the highest form of thinking for which humans are capable?

The answer is like mathematics itself. New theorems are constructed from earlier theorems which are themselves based on even earlier theorems, etc. New thoughts start with earlier thoughts, etc. Memory is essential to sustain the chain of thoughts.

Starting from an elementary beginning, one idea leads to another and the result is today's mathematics, indispensable to modern science and technology and in everyday life.

24.15 Summary

All animals think. They move to satisfy basic needs (food, water, shelter, companionship, intimacy, family unit, community, etc.) to deal with predators and the elements. Hunger and fear drive their actions. Thought and reasoning are qualities essential for survival. Human ancestors had the same motivation.

Having capacity for producing sounds associated with objects found in their environment, language developed. This facilitated interpersonal communication and the ability to record thoughts in memory.

Cognitive memory plays a crucial role to the development of thought and to the retention of thought.

Manipulating objects in the environment helps to create more objects to think about. Thus, thought builds upon thought. Modern thought can be traced back to the thoughts of human ancestors of many millions of years ago.

Appendix

Use google (or your favorite internet search engine) to explore:

- Animal thinking
- Origins of human speech
- Origins of human tool making
- History of hunting
- Origins of agriculture

References

1. University of Cambridge: Inclusive design toolkit. https://www.inclusivedesigntoolkit.com/
2. Widrow, B.: Cybernetics 2.0: A General Theory of Adaptivity and Homeostasis in the Brain and in the Body. Springer, Berlin (2022)
3. Hebb, D.O.: The Organization of Behavior: A Neuropsychological Theory. Psychology Press, London (2005)
4. Isaacson, W.: Leonardo Da Vinci : The Biography, p. 2018. Simon & Schuster, London (2018)
5. Zimmer, C.: Discovering the first known Neanderthal Family. New York Times Science Times, p. 3 (1 Nov 2022)
6. Zimmer, C.: Whales pick up catchy tune when they hear it. New York Times, p. 3 (6 Sept 2022)
7. Incorvavia, D.: How well can you decipher ape-speak? New York Times, p. 8 (31 Jan 2023)

Chapter 25
The Meaning of Meaning

Abstract When reading a paragraph of a book or a newspaper, one can extract the meaning of the content by forming a mental picture of the collection of words and searching long-term memory for a previously recorded mental image to which it is related. Connecting the old with the new brings the new into an established reference frame. The meaning or an understanding of the new is established by exploring the memory and finding a connected image that had been previously studied and recorded. If there are no search hit results, then the new image is recorded and becomes part of the memory's database. Creating mental images from reading text or from listening to music is an important element of meaning abstraction. It is a form of data compression. It is elemental to recognition of complex scenes or sounds. Feature extraction is the key.

25.1 Definition of Meaning

Searching for a definition of the word "meaning" does not result in clarity, i.e. a simple definition. The meaning of this word is taught by examples of paragraphs taken from the literature. There are two points of view: (1) that of the transmitter and (2) that of the receiver. The transmitter is a person who conveys information to others by spoken word or by actions. The receiver is a person who derives information from the environment or from another person. Merriam-Webster defines the meaning of the word "meaning" by giving examples of paragraphs both from the point of view of the transmitter and the receiver, but it does not give a simple definition of the word [1, 2].

This chapter presents a definition of the word "meaning," but from the point of view of the receiver. The following is a definition for the word:

> Sensory information currently received has meaning to the individual if there is some information stored in memory that it corresponds with.

Obtaining meaning is basically pattern recognition. There are other ways to define the word "meaning," but this definition represents the views of these authors. There is no other "official" definition of the word. So, according to these authors, finding

B. Widrow and E. P. Katz, *Cognitive Memory*, Springer Series on Bio- and Neurosystems 17, https://doi.org/10.1007/978-3-031-80939-2_25

meaning is performing pattern recognition. If there is a hit or multiple hits in the cognitive memory, meaning is established. If there is not a hit, and the incoming information is interesting, it will be added to the database. Cognitive memory is the key element in meaning extraction.

This chapter presents many examples of meaning extraction taken from everyday life. As this material is read, the reader should keep in mind the above definition of the word "meaning".

25.2 Meaning as a Mental Image

When you read a complicated paragraph, and you fight to expose the essential meaning, you are creating a mental image of the events of the paragraph, or the thoughts described in the paragraph. You are fighting to understand possibly multiple things. The mental image prompts the memory, relating what you have just read with previously experienced events, thoughts, images, etc. You are bringing the new information into the reference frame of previous information. Sometimes this is difficult, and you need to fight hard to develop new images of what you have just read based upon images from previous experiences.

When fighting hard, the memory is being searched, seeking a hit. If successful, the connection between the old and the new is made. If not successful, the prompt is new information and is added to the database.

25.3 Mental Images from Cognitive Memory

Establishing the meaning or understanding of a paragraph is the process of connecting the new to the old. This takes place within the cognitive memory in humans and in other animals. Understanding something comes from relating it to previously stored knowledge. This is "*meaning extraction.*"

A baby or a young child often does not make such connections because there is little or no previously stored information with which to connect. Thus, they often fail to extract the meaning of a complicated paragraph. Inexperienced young people can sometimes do foolish or dangerous things because even though they may recognize the foolish situation, they may not recognize or understand the extent of the consequences of such foolish actions.

25.4 Visual and Auditory Meaning

Extracting information from either viewing a physical scene or a photograph is a process that is analogous to that of extraction from a written paragraph.

An interesting question is, "How does one extract information from an abstract work of art, from a ballet, or from symphonic music?" The first time you hear a novel musical form, such as jazz, you might not know what to make of it and possibly won't like it. After hearing and/or seeing several jazz performances, you might begin to appreciate what it is all about, and you might even begin to appreciate it. Young people brought up listening to rock music will find themselves in strange company when attending their first symphonic concert. There would be nothing in memory for which to relate this, and it would be hard to derive meaning from this experience.

25.5 The Meaning of Beauty

> Beauty is in the eye of the beholder.
> —anonymous, third century BC in Greek

What is beautiful depends on the experience of the beholder. One can develop a taste for an art form by continued exposure. The first time that you saw a work of art by Picasso, what did you think of it? After seeing a dozen of his works, you will understand what the artist was expressing. But you might or might not like it.

25.6 A Professor Widrow Story, Love of Music

> During the Second World War, I was a teenager. At that time, civilian products were not being manufactured as the total effort of American industry was dedicated to war production. As soon as the war was over, civilian products came onto the market. FM radio became popular. The beginnings of TV broadcasting gave us black-and-white images on tiny screens. My mother bought a radio-phonograph console, and this fascinated me. The phonograph had an automatic record changer. (Wow!) She acquired a nice collection of 78 RPM records: Johann Strauss, Jr. and Frans Lehár Viennese waltzes, and Chopin mazurkas and waltzes. I never before listened to music, but the technology caught my attention. When I heard Strauss's "The Beautiful Blue Danube," I was hooked. I told some of my high school classmates and one of them gave her opinion, "These works are sugar-coated trivia." Undaunted, I loved the music of Johann Strauss, Jr. Listening created an educated taste that I have always kept. Every New Year's Eve, I watch the PBS broadcasts of Great Performances from the Musikverein Hall in Vienna, Austria. The beautifully dressed audience, the Vienna Philharmonic Orchestra playing Johann Strauss and Franz Lehár waltzes, and the accompanying dancing by the Vienna State Ballet, captivates me.

25.7 Beautiful Music

What makes music beautiful? What is its meaning? How does the composer connect with people? What is being expressed? What is the gist of a piece of music?

Music of a given composer becomes likeable and understandable when you can identify this composer's work when listening to a piece which you have never heard previously. Among my favorite composers are Antonin Dvorak, Mahler, Mozart, Beethoven, Stravinsky, Tchaikovsky, Rachmaninoff, Shostakovich, and Aron Copland. This list is much longer. Their works stand out and are recognizable.

25.8 Another Professor Widrow Story: Sculpture

The late Albert Elsen, Professor of Art History at Stanford University, was a close personal friend. When together, we always talked about art. One time the subject of our discussion was about the British sculptor, Henry Moore. I had admired his sculpture outdoors at the Davies Symphony Hall in San Francisco. Each example is a strong work that says something. Just what, I was not sure. I asked Professor Elsen, "What makes Henry Moore–Henry Moore?" He emphatically replied, "Henry Moore worked his ass off." I then asked him, "What do you mean?" The Professor went on to describe Henry Moore as a prodigious worker who expressed something special with each sculpture. After seeing a few of Moore's pieces, you could easily identify a new sculpture of work as belonging to him. There is a strong consistent message coming from each sculpture. And he produced a considerable number. Many of Henry Moore's sculptures can be seen in prominent outdoor locations in various cities worldwide.

25.9 Rodin

Professor Elsen was the world's premier scholar of Rodin. At the Rodin museum in Paris, his book on Rodin was on sale in both French and English. The Professor spent a sabbatical year in Paris, most of the time at Rodin's house. The third floor of which was closed after Rodin's death, but it was opened for Elsen. Inside he found moulds for sculptures that had never been committed to bronze. What he found was an amazing trove of work, finished or not, covered with dust and cobwebs. Left as Rodin had left it.

Professor Elsen became friends with a New York financier, B. Gerald Cantor, who lived in Los Angeles, and who had been collecting Rodin pieces for many years. Elsen convinced Mr. Cantor to donate the bulk of his collection to Stanford University. The objective was to create the Cantor Museum at Stanford that would create one of the finest collections of Rodin art in the world. This museum now contains 200 Rodin sculptures. They decided to create an outdoor sculpture garden resembling Rodin's Garden at his house in Paris. Several works were not in Mr. Cantor's collection that Elsen and Cantor both wanted for the garden. Mr. Cantor was able to purchase some of these on the open market. But others were just not available. However, Elsen knew that the moulds still existed. The prize would be Rodin's "The Gates of Hell." Elsen was able to convince the French government and the descendents of Rodin that Stanford and Mr. Cantor were serious about making a museum for Rodin's work at Stanford. And that it would be the finest in the world, except for the Rodin Museum collection in Paris. Permission was finally given to cast the needed sculptures from Rodin's moulds.

Professor Elsen chose Fonderie Alex Rudier, Rodin's original foundry, to make a new "Gates of Hell." Every other month, he flew to Paris to advise and supervise the sculpture casting, which consisted of 25 pieces that would be carefully welded together in the finishing process. The work, after being shipped to the U.S., spent some time at the Smithsonian in Washington, D.C. while the Stanford University Museum Garden was being prepared. It was finally shipped to Stanford and installed in a manner like Rodin's Garden in Paris. It can be seen today and is the centerpiece (Stanford's sculpture is number 5). There are only five examples of this sculpture in the whole world; this is probably the finest work ever created by Rodin.

Rodin's sculptures are strong, and they speak to you. When viewing a Rodin sculpture, there is no doubt that this truly is Rodin's work. Professor Elsen said that a fine piece of work may not be pretty, but it is important in the history of the artist's work. It is this history that contains the essence and the meaning of the artist's work.

Most of Picasso's work is not pretty. But it is strong. After seeing some of Picasso's works, it grows on you. You can remember it, and perhaps one can find a message with meaning.

25.10 The Meaning of Facial Expressions

Human facial expressions convey meaning. They reflect on an individual's mood or state of mind. The effect is so strong that it can be represented by something simple and elemental such as emoji's. Typical expressions are sad, happy, nervous, joy, surprise, angry, etc.

To have a sense of meaning from facial expressions, it is instructive to follow facial perception as a baby develops from birth. A baby model is a good one for this study because capabilities are added one at a time and this makes it easy to discern the various aspects of meaning of facial expressions and the effects they have on a baby.

25.11 Creation of Beauty

There was a recent documentary video about the life and times of the violinist, Itzsak Pearlman, who is the finest violinist in the world still performing today. The documentary describes his early life as a young man puzzled about how some violinists, who have a gift of being able to produce nice sound, but not of the greatest beauty. What he enjoyed most was teaching violin playing to young people. He was able to sense giftedness in the very few.

The video shows him later in life having gained an understanding of giftedness. He talks about his own experience. He said that he listens to his own playing and practices over again until the sound is beautiful to him. He can sense beauty in his own music that is his key to producing beautiful sound. This illustrates a form of reinforcement learning where practice connects the mechanical motion of the bow and the fingering to the creation of beautiful sound. The feedback comes from listening to his own music, and the quality of the sound depends on his sense of beauty. Fortunately, most people who hear him agree with his sense of beauty. In 2015, he received the Medal of Freedom from President Obama in a ceremony at the White House.

Mrs. Widrow enjoys doing needlepoint. She acquires canvases and colored yarn from designers in England and California. Following a pattern on the canvas, she is able to perform the tedious needle work to create a beautiful image. From hundreds of candidate canvases, she will choose the one that she likes and is committed to expend the required effort to create what she considers to be beautiful. Her sense of beauty is the driving force of her work. An innate sense of beauty drives creative work.

There is an analogy between making music and doing needlepoint. The musician follows the notes and directions of the composer. The art is in the interpretation. With needlepoint, an artist draws the original image on the canvas. Doing needle point, there are myriad decisions involving color and type of stitch. The art is in the interpretation. What meaning is conveyed when one senses beauty?

25.12 Learning Mathematics

Here is another Bernard Widrow story:

> As a first-year student at MIT, I took five different courses, one being calculus. Having studied differential calculus in high school, this was an enjoyable subject. At one point, we were studying multivariable functions, and their partial derivatives. We were taught that the partial derivative of f(a, b) with respect to a and then with respect to b is equal to the partial derivative of f(a, b) with respect to b and then with respect to a. I was never satisfied with simply accepting what I was taught. I spent the entire weekend pacing back and forth trying to prove to myself the veracity of the partial differentiation methodology. No proof or intuitive explanation was given in class.
>
> On Sunday evening, I could draw a mental picture of a two-dimensional function and its derivatives. I fought hard and was finally able to reconcile the new knowledge with previously learned theorems and constructions.
>
> All through my undergraduate years, I fought against the acquisition of new knowledge until in each case I was able to bring this into my own reference frame and thereby gain an understanding. My close friend, Mark Beran, told me in my senior year that I was a fool. On a test, I was spending half the time deriving the equations. Mark said memorize the formulas and use them, nevermind understanding. When taking a test, I would bring my test score back to Mark. My grade would be 98%. Mark would say, "Don't let it happen again!".
>
> I followed his advice. My grades shot up. With rising grades, I was able to get an assistantship from the Digital Computer Laboratory. I worked on magnetic core memory. I was finally off my father's payroll!

Mathematics is a language that helps organize abstract thoughts [3]. When proving a theorem, each next step depends on pulling existing mathematics from memory. Establishing connections gives meaning to mathematics. A proof is a chain of connections. Each step builds on previous steps. Having cognitive memory is essential.

25.13 Mental Images

As a person lives life, mental images are created constantly. If interesting, they will be recorded in long-term memory. These mental images are what is recalled when thinking of objects, sounds, and events of long ago.

The subject of mental images brings to mind a poem, circa 1825, by Samuel Woodworth entitled, "The Old Oaken Bucket." The following is the first stanza of the poem:

How dear to this heart are the scenes of my childhood,
When fond recollection presents them to view.
The orchard, the meadow, the deep-tangled wild-wood,
and every loved spot which my infancy knew.

25.14 Summary

Walking downtown in a city, one encounters people, cars, shops, restaurants, etc. Each scene yields visual images, auditory images, etc. What meaning does each image convey? If you were a person from another planet observing these scenes for the first time, what would all this mean to you? If you were born and brought up in an indigenous tribe in the Brazilian jungle, what would you make of city life seen for the first time? If you were a city dweller walking in a forest for the first time, you would see trees, shrubs, and grass and this would make you happy. You also might encounter a snake or a puma that would cause fear and panic. The meaning of the images is clearly based upon prior experiences and innate knowledge. All meaning is based on comparison of present images with previously recorded images. Arriving from another planet or from the jungles of Brazil, the present scene would be baffling as there would be nothing in memory for which to relate. Memory search would yield no hit. If there were a hit, a wealth of information would be available to assess a new scene, the entire contents of a hit folder. Meaning and understanding will have been extracted.

Appendix

Use google (or your favorite internet search engine) to explore:

- The meaning of meaning
- Wikipedia - The meaning of meaning

References

1. Meaning. Merriam-Webster (2011). https://www.merriam-webster.com. Accessed 8 May 2011
2. Ogden, C.K., Richards, I.A.: The meaning of meaning. Lang. Sci. **41**(Part B), 212–221 (2014). https://doi.org/10.1016/j.langsci.2013.10.001
3. Chomsky, N.: New horizons in the study of language and mind. Cambridge University Press, Cambridge (2000)

Appendix
Compendium of Photographs

A large collection of photographs exists in Professor Bernie Widrow's office at Stanford University. These photographs document some of the early history of the development of neural networks and learning algorithms. As Professor Widrow is now 94 years old and has difficulty in retrieving these images, he asked Professor Greg Kovacs to go to his office and bring appropriate photographs to the light of day.

After reviewing these photographs, this Appendix was planned and expected to be of interest to the reader of this book. The photographs were curated by Profs. Widrow and Kovacs and the result is seen here. It is their hope that you enjoy them (Figs. A.1, A.2, A.3, A.4, A.5, A.6, A.7, A.8, A.9, A.10, A.11, A.12, A.13, A.14, A.15, A.16 and A.17).

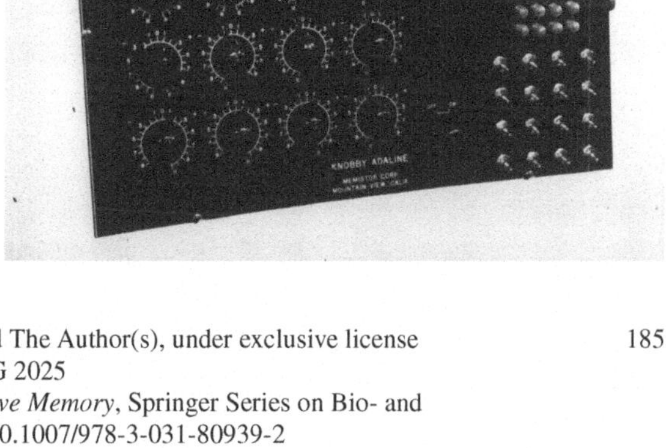

Fig. A.1 Early Knobby Adaline with training with indicator lamps, built at Memistor Corporation c1960. This was a physical incarnation of the initial concept: a mechanically-adapted neuron driven by a human being's hands

B. Widrow and E. P. Katz, *Cognitive Memory*, Springer Series on Bio- and Neurosystems 17, https://doi.org/10.1007/978-3-031-80939-2

Fig. A.2 Bernie Widrow explaining the Adaline concept in Leuven, Belgium, c1967

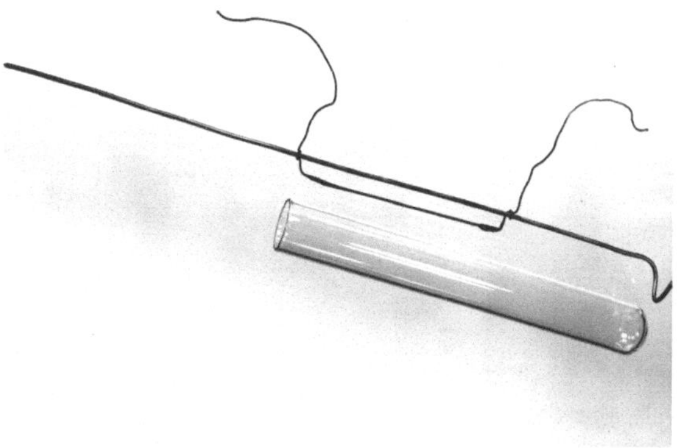

Fig. A.3 The first version, or "test-tube" Memistor c1959. A piece of pencil lead formed the variable resistance, which could be adjusted by plating and de-plating copper onto the lead using a third electrode (thicker metal rod). Later, using specialized resistive wire in place of the pencil lead, miniaturized Memistors were developed that fit within then-standard transistor packages. This was the first step toward an all-electronic neuron that could be adapted using human input or with entirely electronic inputs

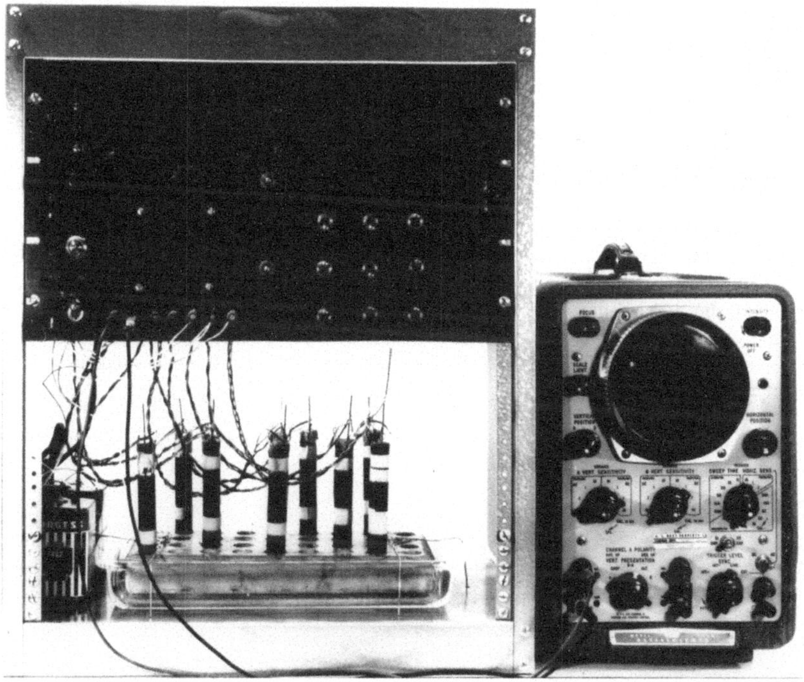

Fig. A.4 The very first Memistor ADALINE, based on test-tube Memistors. This demonstration system was shown on television on "Science In Action" in 1963, but was fully detailed in Technical Report No. 1553-2, dated 10/17/60, sponsored jointly by the U.S. Air Force and U.S. Navy (ONR) available at: https://www-isl.stanford.edu/~widrow/papers/t1960anadaptive.pdf

Fig. A.5 Still frames from "Science In Action" television show in 1963 showing, clockwise from top, Bernie Widrow and host Dr. Earl Herald, the mechanically adapted Knobby ADALINE, the original "test-tube," electronically adapted Memistor Adaline that would later be automatically adapted (no human), the first Memistor ADALINE, then-student J. S. Koford demonstrating Memistor-based speech recognition, and Bernie Widrow summarizing his thoughts (https://www.youtube.com/watch?v=IEFRtz68m-8). The reader is encouraged to listen to Bernie's predictions of the future at about 20:48 in the video

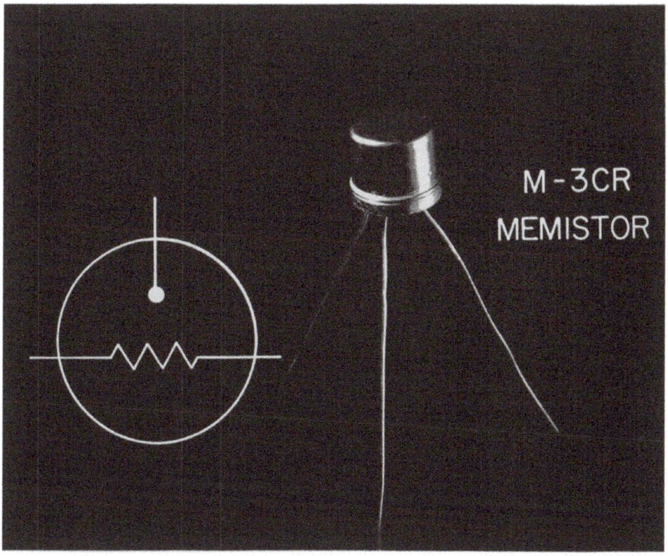

Fig. A.6 Type M-3CR Memistor in transistor-sized (TO-5) package

Fig. A.7 Internal glass envelope of miniaturized Memistor, showing fine resistive wire between left and right leads, control electrode on the center lead, and copper electroplating solution. Once tested, a metal can was sealed to the base, finishing the Memistor. Photo by Gregory Kovacs, 2024

Fig. A.8 A Memistor Adaline, based on miniaturized Memistors, c1961

Fig. A.9 Electronics board from Memistor Adaline c1961. Sixteen Memistors can been seen by their distinctive packaging label (Memistor symbol) as well as sixteen 2N404 PNP Germanium transistors

Fig. A.10 Studio portrait of Bernie Widrow c1963

Fig. A.11 Bernie Widrow working on spatial noise cancellation algorithms in his laboratory with student John Kaunitz in the basement of the Durand Building, Stanford University, c1972. The algorithms being tested later proved to be unworkable, with success coming later with the "Filtered-LMS" approach

Fig. A.12 Polaroid snapshots taken by Bernie Widrow of his EE280 class presentations at his Stanford home, June 3, 1972. Left-to-right, top-to-bottom: poolside discussions, including a young Professor Marty Hellman at lower right; student presenting a project; view from Bernie's study with his MIT chair at lower left; view of students listening; volleyball, with Bernie's daughter Debbie; and Debbie taking the plunge into the pool

Fig. A.13 Bernie Widrow on December 10th, 1995, sitting in the giant chair made for President Taft in 1909. Taft did not appreciate that the chair was appropriately sized for his rather large girth, and forbade the taking of any photographs of him sitting in it. The chair is in the lobby of the Mission Inn Hotel in Riverside, California

Fig. A.14 Bernie Widrow on April 23rd, 2015 at an event honoring his former student, Marcian "Ted" Hoff, designer of the first single-chip microprocessor, the Intel 4004. Left-to-right, Clark Kovacs, Reid Kovacs (currently working toward his Ph.D. in EE at MIT, walking in Bernie's footsteps), Bernie Widrow, and Ted Hoff

Fig. A.15 Bernie Widrow in his home study on Stanford campus on May 17, 2019, demonstrating the Memistor Adaline as seen on "Science In Action," still working nearly six decades after being built. Notably, Bernie's own neurons were still working nine decades after their debut

Fig. A.16 Bernie Widrow on December 30th, 2019, in his Stanford home celebrating his 90th birthday (which was December 24th, 2019) with chocolate cake. Left-to-right, Reid Kovacs, Bernie Widrow, Ronna Widrow

Fig. A.17 Bernie Widrow in Redwood City, California, on December 7th, 2023, prior to his video interview for the Computer History Museum, to document his contributions to artificial neural networks

Epilog

Based upon living experiences, we have explored various questions about human memory. Giving thought to this, we have developed a model for certain aspects of human memory. This model can be simulated or physically built using conventional computer chips and circuits. The purpose of this is two fold: (1) to perform basic science on the one hand and (2) develop a design for a cognitive memory system for modern digital computers.

Such a memory could work in conjunction with conventional computer RAM, giving the computer additional memory capability. Having a human-like memory, the computer would be capable of efficiently solving problems in AI, pattern recognition, and control systems. On the science side, controversial ideas are presented that memory is stored in the brain's DNA, but not in the brain's neural networks. The purpose of the neural networks is for memory retrieval, not memory storage.

B. Widrow and E. P. Katz, *Cognitive Memory*, Springer Series on Bio- and Neurosystems 17, https://doi.org/10.1007/978-3-031-80939-2

The manufacturer's authorised representative in the EU is Springer Nature Customer Service Centre GmbH, Europaplatz 3, 69115 Heidelberg, Germany. If you have any concerns regarding our products, please contact ProductSafety@springernature.com

Printed and bound by CPI Group (UK) Ltd, Croydon, CR0 4YY
01/10/2026
02225941-0002